Writing and Managing
SOPs for GCP

WRITING AND MANAGING
SOPs — FOR — GCP

SUSANNE PROKSCHA

CRC Press
Taylor & Francis Group
Boca Raton London New York

CRC Press is an imprint of the
Taylor & Francis Group, an **informa** business

CRC Press
Taylor & Francis Group
6000 Broken Sound Parkway NW, Suite 300
Boca Raton, FL 33487-2742

© 2016 by Taylor & Francis Group, LLC
CRC Press is an imprint of Taylor & Francis Group, an Informa business

No claim to original U.S. Government works

Printed on acid-free paper
Version Date: 20150708

International Standard Book Number-13: 978-1-4822-3935-5 (Hardback)

Library of Congress Cataloging-in-Publication Data

Prokscha, Susanne, author.
 Writing and managing SOPs for GCP / Susanne Prokscha.
 p. ; cm.
 Includes bibliographical references and index.
 ISBN 978-1-4822-3935-5 (hardcover : alk. paper)
 I. Title.
 [DNLM: 1. Practice Management--standards. 2. Documentation--methods. 3. Writing. W 80]

 RA971.3
 362.1068--dc23   2015026249

Visit the Taylor & Francis Web site at
http://www.taylorandfrancis.com

and the CRC Press Web site at
http://www.crcpress.com

Dedication

For my mother, Barbara Prokscha.
Mutti, this one is for you!

Contents

Preface

When I first started working closely on SOPs and controlled documents, I could not find any books to bring me quickly up to speed. There were books that listed SOPs for good clinical practice (GCP) with some initial content or discussion on each and books that discussed how to create standard business procedures that were not specific to the clinical trials environment. I found some helpful articles, but they were all limited to a few topics such as why we have SOPs and what makes a good SOP. There were seminars offered by industry organizations that focused on effective SOP writing. Those seminars did help, but I learned the most by doing the work and asking questions. I was intrigued enough by the nitty-gritty details of SOPs and controlled document management that I soon found myself taking notes of examples for good and bad practices. As those examples began to accumulate, the idea of writing a book began to form. After several years of mulling over the idea, I decided to write this book to fill the gap in available material and provide introductory information on SOPs and the other documents that relate to SOPs specifically for the GCP environment.

Those who are familiar with my previous book, *Practical Guide to Clinical Data Management* (CRC Press), may wonder how I got from there to here. In 2008, I was hired for a position in clinical data management (CDM) that gave me the opportunity to focus on SOPs, supporting documents, and training rather than on study work. With my entire focus on these needs, and working closely with the Controlled Document group, I came to understand in detail the decisions and circumstances that caused problems for, and noncompliance by, users. Since then, I have worked at or consulted with several additional CDM and Biometrics groups that were also dedicated to processes, standards, and training. As a CDM and Biometrics representative, I participated in working groups for many cross-functional SOPs and also reviewed SOPs from other functions in Clinical Development to assess their impact on Biometrics. My work also included participating in cross-functional initiatives whose charter was to implement electronic trial master file (TMF) systems, and those experiences also show up in several chapters.

The approach in this book is unique in that I come at the topic as a user, reviewer, and author of SOPs, rather than as a member of a regulatory or quality group. Because I was very close to the actual processes documented in and governed by SOPs, I saw how noncompliance occurs—in fact, I was the person responsible for filing the required paperwork on those occasions. These experiences allowed me to see what can be done to better write and manage SOPs to improve compliance and also support staff in their work. The idea of specifically calling out approaches to SOP creation and maintenance to make it easier for users to stay in compliance is a theme found throughout the chapters of this book. Those examples that I have been noting over the years are here as well; though the companies are not identified, the examples are accurate reflections of real-world experiences.

When I first conceived of this book, I thought of its title as *An Opinionated Guide to SOPs*, and though I changed the title for publication the underlying idea that the recommendations are my opinion on what should be done has remained. I believe that expressing a preference will serve the reader more than presenting all the options equally. I am also not shy of going a whole new way when the old way has limitations. I have been very fortunate to have had the opportunity to try out and refine many of my recommendations at biopharmaceutical companies prior to, and even during, the writing of this book. What I have learned is here for your benefit, illustrated by scenarios I encountered along the way.

Susanne Prokscha

Acknowledgments

My special thanks to Rena Liu-Critchley, who was instrumental in convincing me that my ideas about SOPs were worth putting down and sending out into the world. Rena provided valuable feedback from the original outline through to submission of the manuscript; if I wanted to discuss controlled document numbering schemes or share examples to see if they were compelling, she was the person I turned to. Rena has been very generous with her time during a busy period in her life and I will always be grateful.

When I was writing tricky sections and wanted to make sure my ideas were clear and reasonable, I often turned to my "test readers," Clarice Grant and Wakana Kim. They frequently heard me say, "I was just working on that question for my book..." and never seemed to get tired of hearing it. Their enthusiasm and support are greatly appreciated.

A final note of thanks to Beth Fallon. It was Beth who first hired me for a position that gave me the opportunity to focus on SOPs, supporting documents, and training rather than on study work. That position allowed me to build on the expertise I already had and take it in a new and satisfying direction. Beth is also a role model, whose style of working and dealing with people I strive to emulate.

About the author

Susanne Prokscha graduated with a degree in physics from MIT. Her early work at a medical device company led to an opportunity to work on software for clinical trials. She has been involved in clinical data management (CDM) processes and technologies since the mid-1980s. She has worked as a consultant and directly for both large and small companies, gaining experience with a wide range of studies and a variety of clinical data management systems. Her interest in advancing the field of CDM and helping emerging CDM groups led her to write *Practical Guide to Clinical Data Management* (CRC Press, Boca Raton, Florida), the third edition of which was published in October 2011.

Since 2007, Susanne has been focusing on process development, SOP writing, document management, and training for CDM and Biometrics groups. She currently works as director in clinical data management for process and training at Onyx Pharmaceuticals (San Francisco, California), an Amgen subsidiary.

section one

Founding principles

chapter one

Introduction to SOPs

Before discussing the details of writing and managing standard operating procedures (SOPs), we will have to define SOP and establish what purpose SOPs serve in biopharmaceutical and device companies.

What is an SOP?

*ICH E6 GCP** defines *SOPs* as "detailed, written instructions to achieve uniformity of the performance of a specific function." Unfortunately, nearly all the individual words in this definition except for *written* allow for enough leeway to make this definition of SOP too vague to be useful to a company trying to create or fine-tune its SOP philosophy. *Detailed* implies something more than a general statement of intent, but how detailed is *detailed*? *Uniformity of performance* is the "standard" part of standard operating procedure; some things need to be done the same way each time and following the SOP should achieve this goal. That does seem pretty clear. *Specific function* is the procedure or activity in question, but the definition does not say what kind of function or procedure or how broad its scope is: Can SOPs for clinical research govern any function or can they govern only regulated activities?

Consider this alternate definition for SOP:

> SOPs are written instructions that identify the activities and responsibilities needed to achieve a standard, controlled procedure that ensures compliance to GCP and applicable regulatory requirements and reflects business needs in support of clinical research.

This language emphasizes that the procedure being documented is already standard and controlled. It also begins to address the level

---

* *ICH E6 GCP* (International Conference on Harmonisation, *Guideline for Good Clinical Practice*, E6) will be used herein to refer to the document. GCP (good clinical practice) alone should be taken to refer more generally to the actual practice, which is defined and guided by a variety of regulations and guidelines issued by health authorities to govern the conduct of clinical trials.

of detail ("activities and responsibilities") and areas SOPs should be applied to ("ensures compliance"). But even this alternate definition leaves open many practical questions of implementation. The chapters that follow use this definition as a guide and address the practical issues in more depth.

Why do we have SOPs?

Every biopharmaceutical company has SOPs, and new staff members are told that the company "has to have them" and that each employee "has to train on them," but rarely is anyone told why there have to be SOPs and even less frequently are they explicitly told that they must follow them as written. So it is perhaps not surprising that the question comes up over and over again of whether SOPs can be done away with, as they often seem to have so little practical value and are a great nuisance to manage. Individual employees sometimes dare to raise this question (and these employees are typically ignored), but upper management also raises the question of whether SOPs are *really* required (and these managers typically must be heard and responded to). SOPs are used in clinical research because they are, in fact, a regulatory requirement but they also can improve business effectiveness by disseminating best practices and creating a contract or service agreement between different parts of the organization.

GCP requires SOPs

Most clinical trials conducted by biopharmaceutical or device companies are conducted in a regulated environment; that is, trials are subject to rules and regulations imposed by the countries in which they are conducted. *ICH E6 GCP* has been adopted as regulation by the ICH member countries: the European Union, Japan, and the United States. Any marketing submission made in those countries must be based on clinical trials conducted according to *ICH E6 GCP* regardless of where those trials were conducted. Other countries have also put into regulation these same guidelines or guidelines that are reasonably similar—so it is safe to say that all organizations should adhere to the requirements of *ICH E6 GCP* even in countries where it is a guideline rather than a regulation.

If a clinical trial is being conducted according to *ICH E6 GCP*, then according to Section 5.1.1, "The sponsor is responsible for implementing and maintaining quality assurance and quality control systems *with written SOPs* to ensure that trials are conducted and data are generated, documented (recorded), and reported in compliance with the protocol, GCP, and the applicable regulatory requirement(s) [emphasis added]." In addition to this broad requirement, additional, more specific, references to SOPs follow, which make clear that SOPs are expected to be in place for all key activities associated with a clinical trial. For monitors, Section 5.18.2 ("Selection and Qualification of Monitors") says, "(c) Monitors should be thoroughly familiar with the investigational product(s), the protocol, written informed consent form and any other written information to be provided to subjects, the sponsor's SOPs, GCP, and the applicable regulatory requirement(s)." And for data management, Section 5.5.3 says, "When using electronic trial data handling and/or remote electronic trial data systems, the sponsor should: ... (b) Maintain SOPs for using these systems." In order to be in compliance with *ICH E6 GCP*, all biopharmaceutical companies must have SOPs for clinical trial activities.

The organization needs them

Organizations also benefit from SOPs beyond meeting regulatory expectations in that they can codify best practices and agreements. Growing companies often find that when they start to run trials in growing companies often find that when they start to run trials in multiple therapeutic areas, their teams begin to run trials differently. New teams may find themselves figuring out the right process as they go. When these companies write or revise SOPs, they can disseminate best practices to all trials to improve efficiency as they grow.

A very important function of SOPs for both large and small organizations is to act as a kind of contract between departments to formally record who is responsible for which tasks. When SOPs are written, one goal should be to ensure that the most appropriate department or group is responsible for completing each activity listed in the procedure. Accepting responsibility for a task implies a commitment of the resources needed, including staff and materials, to carry out that task consistently. Senior management from each department involved in the procedure will approve the SOP, which implies agreement to commit the resources, and all groups will train on the SOP, ensuring that everyone is aware of the agreements regarding responsibility. In cases where this agreement is *not* formalized in an SOP, the commitment to allocating appropriate resources can be lost and may even be challenged. Note that the "service agreements" used by some companies for commitments of responsibility across groups do not come with training requirements and are often not known to the staff performing the work, as seen in Example 1.

Example 1

At one midsized company, the Information Technology (IT) group provided systems support for the electronic data capture (EDC) system used for clinical databases, including supporting software integration of EDC with other systems. During a revision of the process, IT representatives and managers agreed to perform certain steps during study setup to initiate and test integrations, in effect developing a service agreement. Because IT staff were not held to, nor were they trained on, Clinical Development SOPs, those not involved in the discussions were not aware of the agreement. As the process was rolled out, IT staff who should have been performing the steps refused to do so, impacting the study database release process.

Why do we follow SOPs?

Even when employees know SOPs are required by GCP and can benefit the company, there is often still resistance to following certain steps in SOPs when those steps are inconvenient, labor intensive, or no longer quite correct. All parts of the organization need to make clear to all employees that regulatory agencies expect SOPs to be followed. The definition of *audit* found in *ICH E6 GCP* reinforces the message that SOPs are essential to running a trial and makes it quite clear that activities during study conduct must adhere not just to regulatory requirements but also to the protocol and SOPs: "Audit: A systematic and independent examination of trial-related activities and documents to determine whether the evaluated trial-related activities were conducted, and the data were recorded, analyzed, and accurately reported according to the *protocol, sponsor's standard operating procedures (SOPs)*, Good Clinical Practice (GCP), and the applicable regulatory requirement(s) [emphasis added]." Also in *ICH E6 GCP* we find the following: "5.20.1 Noncompliance with the protocol, SOPs, GCP, and/or applicable regulatory requirement(s) by an investigator/institution, or by member(s) of the sponsor's staff should lead to prompt action by the sponsor to secure compliance."

In some cases, such as when a process has drifted significantly from that documented in the SOP, it would be better to have no SOP than to have serious deviations from an effective SOP, as seen in Example 2. SOPs show that there is a controlled process in place for a regulated activity, but it is still possible to have a controlled process without having an SOP. For example, documenting the processes actually being followed in a study-specific document such as the trial monitoring plan or a clinical data management plan can provide the necessary evidence to an inspector that the protocol, GCP, and other regulatory requirements were being met.

Example 2

Companies can find themselves in a bind with resources in their Controlled Document groups. It is surprisingly common for a Controlled Document group to limit the number of SOPs that can be added or revised over a period of time because of lack of staff to process those documents. One such company found itself in a position of having to retire its SOP on study database build when, after a process reengineering effort, the existing SOP could not be updated but the improved process was so much more efficient that the business needed to adopt it. The company used lower level and supporting documents to demonstrate that the process was indeed in control and updated all training to reflect the new process.

Beyond SOPs

We also have some evidence that *any* formal written and approved procedure in effect at a company must be followed. For the FDA, we can refer to the Compliance Program Guidance Manuals (CPGMs), which the FDA uses to direct its field personnel on the conduct of inspectional and investigational activities. In CPGM 7348.810, titled *Sponsors, Contract Research Organizations and Monitors*, there are several references to written procedures other than SOPs. For example, in Section III. G.1.b we read the following instruction to inspectors as they review monitoring procedures: "*Obtain* a copy of the sponsor's/CRO's/monitor's written procedures (SOPs and guidelines) for monitoring and *determine* if the procedures were followed for the selected study. In the absence of written procedures, conduct interviews of the monitors as feasible and/or otherwise *determine* how monitoring was conducted." Similarly the MHRA*'s *Good Clinical Practice Guide*, Annex 1, Section A.1.6.1, informs organizations being inspected that they may have to provide documents including "written procedural documents (for example, standard operating procedures (SOPs), working instructions)" to the inspector. The FDA's inclusion of guidelines and the MHRA's inclusion of working instructions warn us that the scope of auditable documents goes beyond SOPs—and presumably, during an inspection, organizations would be held to the procedures found in those working instructions or guidelines as they would to SOPs.

Say what we do, do what we say

All of the chapters that follow aim to help organizations create SOPs and other associated documents, such as working instructions, in such a way that each one accurately reflects the procedures the organizations

---

* Medicines and Healthcare Products Regulatory Agency, the regulatory agency of the United Kingdom for medicines and medical devices.

intend to follow. Just as important, we will also look at ways to create and manage these documents so that staff can follow them, not only when they are first rolled out but also after time has passed and the regulatory or organizational environment has changed. The goal is—to use a common industry phrase—to say *what we do and do what we say.*

chapter two

Document hierarchies

In Chapter 1, we identified the need for SOPs as part of quality control systems required by *ICH E6 GCP*. SOPs, however, do not stand alone; even at the smallest of companies, activities described in SOPs require the support provided by associated forms, templates, and more detailed work instructions. SOPs and the other supporting document types are often illustrated in the documentation describing a company's quality system (the quality manual) by a pyramid showing the document hierarchy. This chapter introduces a novel version of the document hierarchy diagram, which will be referenced throughout this book. The simple hierarchy shown in Figure 2.1 and the more complex hierarchy in Figure 2.2 are tailored specifically to clinical development activities and show the types of documents linked closely to SOPs. The document hierarchy uses the concepts of *controlled documents* and *managed documents*, which we will explore first.

Controlled documents and managed documents

People who specialize in quality systems often refer to ISO 9000,* which addresses quality management generally, or ISO 9001, which sets out requirements for a quality management system. In ISO 9001, the topic of document control appears as a key element of quality systems because good document control ensures that everyone has access to the right documentation to perform their work. Document control involves the following:

- Formally approving documents to indicate the document is fit for purpose; approval also demonstrates agreement of the parties involved
- Periodically reviewing documents to see if they are still valid
- Updating and reapproving documents as needed, obsoleting them when appropriate, and ensuring outdated versions are not inadvertently used

---

* ISO is the International Organization for Standardization (www.iso.org). ISO 9001 here refers to ISO 9001:2008.

- Maintaining a revision history with version numbers to provide an overview of changes with time
- Making the documents readily identifiable and available for use by anyone who needs them (including external vendors)

References to controlled documents in the chapters that follow imply that these controls are in place.

What documents must be controlled? In the good clinical practice (GCP) world, documents that ensure compliance with regulations and align processes with regulatory agency guidance documents are controlled. SOPs and their very closely related supporting documents are always controlled and are maintained in a validated, *21 CFR Part 11*-compliant* software system. A group within the Regulatory Compliance or Quality Assurance departments generally oversees the procedures to control documents and the software systems used to implement the controls—this group will be referred to as the Controlled Document group in the chapters that follow.

Although controlling documents is generally a good thing, control imposes a formal procedure on the documents and that formality introduces a significant time element to their creation and revision. (See Chapter 10 for examples) In mid- and large-size companies, departments and subgroups will create additional documents to aid in the spread of information and use of best practices in support of the requirements found in SOPs. Staff members of the departments will manage these documents and may put into place controls that meet many—but generally not all—of the same requirements as for controlled documents. Often, the documents are not officially approved by management, although thorough review of the contents is common. Also common is the practice of maintaining and making these documents available on a "shared drive" or internal web page, though it will be read-only and historical versions will be maintained. These kinds of documents will be called *department-managed* documents in the chapters that follow; there is no consistent term already in use in the industry.

Controlled documents have a very strong advantage over department-managed documents that is not apparent from the bulleted list above because it comes from the software systems in which controlled documents are managed. Most controlled document systems have built-in training features that track who is required to train on controlled documents and whether those people have taken the training. The controlled document system will also automatically reassign training when documents are updated (see also Chapter 15). As noted above, department-managed documents are typically not stored in the controlled document system.

---

* Code of Federal Regulations Title 21, Part 11.

Therefore, the automatic tracking of training does not take place. If the material in a department-managed document is essential to performing a task, that material must be included as part of other training, or training on the document alone must be required and tracked.

Overview of the hierarchy

Figure 2.1 shows the document hierarchy that would apply to Clinical Development in every company. The top triangle in the pyramid is a company's Quality Manual, which describes, among other things, the hierarchy and how the types of documents are to be managed. This explanation of the management of documents may instead (or in addition) be written in the form of an "SOP of SOPs" (see below). The second level of the pyramid has SOPs in the middle, strongly impacted by the SOP of SOPs above. Supporting the SOPs on either side are closely associated *forms* and *templates* on one side and detailed *work instructions* and *manuals* on the other. Documents in these two levels of the pyramid are key components of any company's quality system and are tightly controlled, generally by a dedicated group, and stored in a computer system specifically designed for controlled documents. These two levels may be the only official types of documents that smaller companies require.

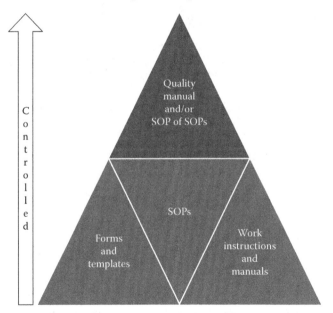

Figure 2.1 A document hierarchy supporting good clinical practice (GCP) at any size company.

In Figure 2.2, we add another level to the bottom of the pyramid with document types that support the needs of mid- to large-size companies. In the new level, we find a number of other document types that also impact the quality and conduct of the company's clinical trials. Two sets of documents: (1) department-managed documents and (2) *training materials and curricula*, directly support the controlled SOPs, forms and templates, and work instructions and manuals. These types of documents, described more below, must always be aligned with the controlled documents and may not conflict with the contents or procedures found there. Documents maintained by departments and training associated with SOPs are both items managed by the departments in Clinical Development rather than by the quality document group (see also Chapter 16).

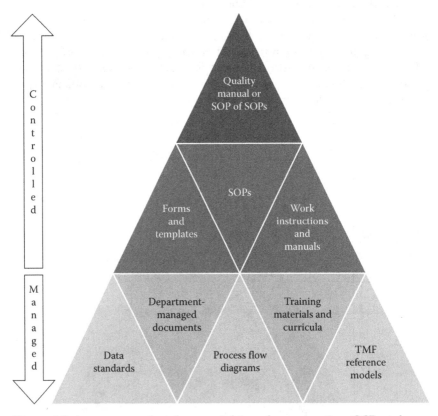

Figure 2.2 A more complex document hierarchy supporting GCP at larger companies.

In the middle of this bottom level, we find *process flow diagrams.* At some companies, the diagrams created during the design phase for a process are managed as stand-alone documents with version control, rather than as a section or appendix of an SOP. These process diagrams may even be approved by management before they are used to create or update SOPs. The link from the diagrams to the SOPs is very strong, but it is not unusual for the staff who work on and manage the diagrams to be different from those working on SOPs so an extra effort may be needed to ensure that the two are always in synch.

In the corners of the pyramid's third level we find two, perhaps surprising, document types that impact SOPs and are impacted *by* SOPs: *data standards* and *trial master file* (TMF) *reference models.* Data standards (including case report form, database, and tabulation standards) are very detailed descriptions used to support consistency in data collection and analysis across studies. They are referred to by SOPs for study startup, data analysis, and creation of the clinical study report. TMF reference models are also used to support consistency across studies—in this case, in the structure and contents of the trial master files associated with those studies. Whenever a controlled or department-managed procedure results in the creation of a study document, the procedure authors will have to assess if and how that document fits into one of the company's TMF reference models. Data standards and TMF reference models are constantly evolving and are generally managed by governance groups or cross-functional committees in Clinical Development. (Data standards are discussed below; Chapter 5 discusses TMF reference models in more detail.)

The further discussion of controlled documents and managed document types that follows clarifies the differences between documents in Levels 1 and 2 and those in Level 3. The next two sections provide enough information for those wishing to get only a quick introduction to document systems; for those reading further in the book, these sections provide the basis for more detailed discussion in the chapters to come.

More about controlled document types

In Chapter 1 and in this chapter, we have established that all companies doing clinical development will have to have a quality system with SOPs maintained as controlled documents. A special SOP will define what "controlled" means for each company. Most companies will also support controlled work instructions and/or manuals to provide a structure for important supporting procedures and information.

SOP of SOPs

ISO 9001 requires a documented procedure to establish the methods to control documents. In the biopharmaceutical industry, information about controlled documents may be found in a quality manual (if there is one) but there is always a standard operating procedure describing how documents are controlled. Although the proper title of the SOP varies (and is usually quite serious), this document is colloquially known as the "SOP of SOPs." The SOP of SOPs starts by defining the document types and their hierarchy or by referring to the company's quality system manual for this information. It then goes on to specify the requirements for initiating, identifying, writing, approving, posting, updating, and retiring SOPs and related controlled documents—covering all the elements recommended by ISO 9001 listed above. To provide all of this information, it is often one of the longer SOPs and it may have several appendices. The process for documenting and resolving planned or unplanned deviations from SOPs needs to be covered also—either in the SOP of SOPs or in a separate SOP.

Although an SOP of SOPs will always cover the controlled documents, it is much less common for companies to define what constitutes department-managed documents and what controls or requirements such management must entail in the SOP of SOPs. Ironically, departments end up developing a department-managed document to cover department-managed documents, leading to a situation where different departments in the same company create such documents and use different definitions and procedures for what is essentially the same type of document. To avoid this situation and establish a baseline for these supporting documents, the SOP of SOPs should at least touch on the topic, as shown in the example SOP of SOPs in Appendix 2.

The types of documents found in the third tier of Figure 2.2 probably never appear in a quality manual or SOP of SOPs. But when these additional documents are not recognized, companies end up missing the connection to SOPs, leading to everything from minor confusion to serious noncompliance.

SOPs

In the first chapter, we redefined SOP as follows:

> Written instructions that identify the activities and responsibilities needed to achieve a standard, controlled procedure that ensures compliance to GCP and applicable regulatory requirements and reflects business needs in support of clinical research.

This definition tells us that an SOP focuses on the list of tasks that need to be completed and the roles or people involved: it documents the *who*, *what*, and *when* of a process. The *how* should only come up if the tool or system is a critical part of the agreed-upon process—as would be the case if a specific form or template must be used. (We will see in Chapter 4 that leaving flexibility in the *how* has the benefit of keeping followers of the process in compliance longer, as methods change over time to improve efficiencies.) Although SOPs should focus on activities around regulated activities (those covered by GCP or other regulations), some activities will be standardized in a company and formalized in an SOP because the organization has a business interest in that activity. Chapter 3 goes into more detail about what should be covered in an SOP and what should not.

SOPs can and often do include tasks across multiple Clinical Development functions or departments. At some companies, if only one department is involved, the document may be called a *departmental operating procedure*. Similarly, some companies with offices geographically removed from each other use different names for SOPs that apply globally and those that apply only to a given region or office. They are all SOPs, but with different scopes of application. The chapters that follow use only the term *SOP*, but this should be read as including these variations.

Forms and templates

Even though SOPs do not generally include the details of how a particular activity is carried out, sometimes a particular tool or system is mentioned because it is key to carrying out the procedures consistently and correctly. *Forms* and *templates* are the most common tools mentioned in SOPs.

Forms are used to provide a highly structured set of information when that information is needed to carry out a process. Forms are generally "short answer," though longer fields for comments, descriptions, and explanations are common. Forms are also used to capture approvals when the work being approved is not a document, as in the case of approving serious adverse event (SAE) reconciliation for a study prior to study database lock.

Templates provide a structure for a document whose content will be specific to a study. Many templates are prefilled with standard content, such as text for protocols or trial monitoring plans or the most typical kinds of data review performed for a particular therapeutic area at the company. Templates frequently have sections that are not to be changed and others that are optional or to be used only if applicable. This is often shown by shaded or colored text or sections and, unlike forms, extensive customization is expected. Forms contain information recording work done (e.g., SAE reconciliation) or they are sent to someone for them to do work (e.g., create an investigator site in the clinical trial management system); in the case of templates, customizing the file contents is the work itself. Although the difference between forms and templates seems clear and intuitive, the vagaries of controlled document systems, as we see in Example 1, can make the line less clear.

Example 1

There are cases where the division into form or template is not quite so clear. Because of limits in their formatting, one company maintained that if there were a list of things to be provided, such as data sets to be included in an interim analysis or the studies to be included in a regulatory submission, the list length—the number of rows or entries—would have to have a fixed limit for it to be a *form*. If it were a list without a preset limit, then it would have to be classified as a *template*.

If a form or template is mentioned in the SOP, does it have to be a controlled document? Although it would seem the obvious answer is yes, and that is the best approach, it may not always be the case. Some documents are hard to manage in the controlled document system because they change frequently. Templates for various study plans, such as those for data management plans, statistical analysis plans, and monitoring plans, have default text that is frequently updated, so these may be managed outside controlled document systems as department-managed documents. Note, however, that when a form or template is in the controlled document system, it must be associated with an existing SOP or work instruction. Many companies require that the SOP or work instruction explicitly lists where in the process the form or template is to be used.

Work instructions and manuals

In laboratories or manufacturing environments, *manuals* or *work instructions* can be as important as SOPs and are carefully maintained and followed. In the GCP environment, the name and intent is a little different, in that these types of documents are provided as supporting documents to procedures or activities that appear in SOPs. Work instructions are more detailed procedures describing the *how* for activities in the SOP, and manuals are more like handbooks, guides, or references. These documents are generally controlled because their content must support SOPs, not contradict them, and because they are essential to performing the work properly. Controlling these documents also ensures that staff are trained on the procedures or information described in them.

At some companies, work instructions are formatted like a procedural document (i.e., SOP) but are permitted to have a great deal more detail. They are typically, but not always, limited in scope to a single function or department. They must be linked to one of the SOPs they support and should explicitly reference that SOP in the *scope* section or *purpose* section of the work instruction. However, it is *not* necessary to mention the work instruction in the SOP. It is a best practice to only create work instructions when they are needed, and a need may not be recognized until after the SOP becomes effective and is in broad use. In fact, some companies maintain that documents should not refer down the document hierarchy, only up or across, to ensure flexibility.

Unlike SOP, the term *work instruction* is not used consistently in the industry. Other terms for this kind of document are *working practice* and *job aid*. Whatever term is used, work instructions are usually easier to manage and update than SOPs; because the application area is usually limited to a single function or department, the review cycle is greatly reduced and the approval is limited to that level also.

The format and content of *manuals* are usually very flexible to permit creation of a document that fits the need in providing information or instructions. Thinking of manuals as handbooks, in contrast to the procedure-focused work instructions, helps define the line between them. One good rule of thumb is that if it has screenshots, it is a manual. Not all companies support manuals as a controlled document type, but there can be value in controlling this kind of nonprocedural document when it applies to critical processes rather than maintaining them as department-managed documents. Note that when manuals *are* supported in the controlled document system some companies do not require them to be associated with an SOP, but as we will see in later chapters, this lack of association can make them harder to locate. In addition to the flexibility in format and content, manuals are even easier to update than work instructions as they typically require only lower management approval.

More about managed document types

Looking now to the bottom tier of Figure 2.2, we see several managed document types that are still closely connected with and impacted by SOPs. These document types are used in Clinical Development by nearly all companies, but smaller companies may be using contract research organizations to perform some of the work conducting trials or may have only a few examples of each type of document, and so they may not see the need to formalize their management. As a company grows and documents multiply, identifying the types and the way they are to be governed becomes a necessity.

Department-managed documents

There are many reasons why departments begin to create documents outside of the controlled document system. In an ideal world, all the documents that a group needs in order to follow consistent and standard processes would be stored as controlled documents. Those documents would be easy to update and require a minimum of red tape to create, modify, and remove. In actuality, getting anything updated once it is stored in the same system as the SOPs can be daunting—leading staff to create uncontrolled documents. Another common reason for not controlling documents is that some companies do not support control of documents other than SOPs and their related forms and templates. When departments at these companies need to disseminate additional information or practices, they are forced to create and manage the documents on their own.

Even at companies with good support in the controlled document system for department documents, there are good reasons to maintain some documents as uncontrolled. In the following cases, department management of the documents makes good sense:

- Situations where the process covered in a document is new to broad use but still needs to be shared
- Documents where frequent updates are always expected
- The need to share informal documents or sample/example documents

Documents in the first category are needed for procedures being formalized for the first time—perhaps to support a new system or process such as more centralized monitoring or moving to the use of electronic patient-reported outcomes. If the process is relatively new but well thought out and piloted (as explained in Chapter 7), it may be desirable to document the process so it can be used by studies that need it and plan on a revision after just a few months. As in Example 2 below, the idea is that the document will be upgraded to a controlled document once the process details have been confirmed.

Example 2

After having prepared several submissions to regulatory agencies, during which problems had been encountered in preparing case report forms (CRFs) for inclusion in those submissions, a company created a working group to formalize the best practices. The working group created, finalized, and posted an SOP that was written at a fairly high level. The Clinical Data Management group decided to create a more detailed work instruction to support the SOP.

> In preparation for the next planned submission, they recorded the more detailed process steps as well as they could and aligned it with the new SOP. The plan was to use the detailed process in the next submission, after which the work instruction would be updated and posted as a controlled document.

An example of the second category, where the documents are expected to be revised frequently, is the *data management plan* template. Because data management plans frequently have references to SOPs and work instructions, vendor-specific data handling requirements, and computer system-specific information, the templates themselves may require updating several times a year. As long as the template adheres to any requirements that might be in the SOP on data management plans, the template can be maintained as a department-managed document. In the third category we find documents such as best practices, FAQs, and example slides for investigator meetings. In larger companies, these informal documents and examples can improve consistency across studies and save employees time—so there is value in making them available more broadly.

As we will see in Chapter 17, managing these documents means imposing some process for review and release to ensure they support, and do not contradict, SOPs and are aligned with existing training.

Training materials and curricula

Whenever an SOP or any other controlled document is released or updated, any training associated with activities covered by that SOP has to be reviewed to ensure that it is aligned with the published procedure. This review is also necessary for training that references posted department-managed documents. Because training materials have such a tight connection to controlled and department-managed documents, as soon as a department has more than just a few training courses (instructor-led or computer-based), the training materials themselves will have to be managed. This becomes even more critical when the trainers are members of the department, whose normal job description does not include "trainer" but who volunteer to give training as a subject matter expert (SME). These SMEs will not be experienced with, nor deeply aware of, the process of preparing and releasing documents and training materials. Someone in the department will need to keep and release versions of training materials and to ensure that they are correctly updated.

In addition to the training materials directly, larger departments will keep a list of what training is required for various roles in the department. This will include SOP training, system or application training, and process training. This list of training by role is often referred to as a *curriculum*. The curricula for the department have to be updated as documents are updated. Chapter 15 addresses the management of training curricula in more detail.

Data standards

Because data standards are described in complex documents and because standards must be followed to ensure quality of the data and to take advantage of efficiencies that standards introduce, some companies will be inclined to control the documents containing the standards as *templates* associated with SOPs. So, for example, the electronic case report form (eCRF) fields and their attributes may be documented in a large Microsoft Excel spreadsheet and controlled as a template in association with the SOP on eCRF creation and release. Problems arise when that spreadsheet has to be revised to correct errors or to add data elements for new types of studies. A controlled template *can* be revised, but there will be multiple revisions in a year and the review cycle normally associated with SOP and other controlled documents does not apply to standards: Most reviewers will not be able to assess if the standard field attributes are correct; the quality is assessed through expert review and creation of the eCRF elements from the specifications. Keeping standards as controlled documents and not updating them can lead to serious compliance issues as shown in Example 3 below.

Because standards are a moving target and because their creation and revision process is completely different from that of a procedural document or associated template, it makes sense to maintain standards as managed documents. Most frequently, there will be a governance group or committee overseeing data standards that will review requests for changes, implement them, and then ensure they go through extensive testing in the applicable system before they are posted.

Example 3

A company that had conducted studies using vendor-created eCRFs in various electronic data capture (EDC) systems decided that all studies going forward would use the same EDC system and that studies would be built in-house. From their experience with the vendor, they were able to create an SOP on eCRF build and release. They created the first set of eCRF standards as a template associated with the study build SOP and posted the set of documents. The eCRF standards very quickly needed revision, but this group never updated the controlled template; all staff were instructed to get the eCRF standard template from a folder on the group's shared drive. In effect, they never used the controlled version of the template. Although an inspector might never notice this subtle noncompliance, it was clearly an area where errors impacting the business could be introduced. A few years later, the standards templates were decoupled from the SOP and moved into a department-managed area. The company also introduced a more formal revision process for the contents of the eCRF standards.

TMF reference models

The same problem that faces data standards posted as controlled templates also applies to TMF reference models. TMF reference models document how essential documents and documents that can be used to reconstruct the conduct of a trial are to be filed. These models may be used for a paper TMF approach, though formal reference models are often introduced with the implementation of an electronic TMF software systems. Many companies may have one model for use for all trials, but some larger companies may use multiple models to support studies in different regions or phases of development. Also, just like standards, staff need to use the latest model for their type of study so there is a strong temptation to control the model with the SOP on TMFs, often as an attachment or appendix.

However, like data standards, TMF models are typically described in very large Excel spreadsheets that undergo frequent revision. The same solution for data standards can also apply to TMF reference models. The SOP on TMFs can refer to the model and the model can be a managed document. A cross-functional working group will assess requests for changes to the filing model and they will have a process in place for updating and releasing new versions.

As we will see in Chapter 5, TMF reference models have a particularly strong link to SOPs because, when a controlled process generates a document, the SOP should say explicitly if that document should be filed in the TMF. If an SOP requires that a document be filed in the TMF, the TMF reference model has to have a clear place to file it!

Process flow diagrams

Many companies have adopted the concept of creating a process diagram before writing a procedural document like an SOP (see Chapter 8). At some of these companies, the process flow diagrams are managed as a specific class of document. They are created, extensively reviewed, approved by management, and released with a version number. After the diagrams are approved, SOPs will be written to cover some of the activities in the diagrams, and those SOPs may have more detailed diagrams that describe more of the prerequisites, hand-offs, and outputs.

Sometimes, writing the SOP and piloting the details will identify issues with the original process flow diagram; at this point the diagrams and SOPs may begin to drift apart. If both are referenced by staff, then they must be kept in synch. At some companies, these diagrams also function as a way of accessing the SOPs written to reflect the process; users click on elements in the diagram to access documents and training (see also Chapter 14).

Other document types

Those familiar with document hierarchies may wonder about some document types that appear to missing.

Where are the policies?

Many document hierarchies used at biopharmaceutical companies place "policy" at the top of the pyramid—so where are the policies in Figure 2.1? The ideal of a policy document is to state the company's intent in any given area; this intent then guides the creation of procedures and related documents for that area. When implemented in this way, the policy provides requirements that the procedures, when written, must adhere to. For example, a policy on monitoring might include a statement such as "Study-specific details for monitoring will be recorded in a trial monitoring plan." The process documented in the SOP on monitoring would then be required to include a trial monitoring plan and to provide the details of who reviews and approves the document and the time point by which it would need to be final.

Unfortunately, and all too often, the policies are written *after* the SOPs or procedures as a kind of abstract or summary of what is found in the SOPs. When polices are written this way, they have no value and are a waste of time to maintain. Even when written before SOPs, the main value of a policy is to writers of SOPs—if staff are required to read the policies at all, they would read them just once and then refer to SOPs to perform real work. Given their limited value, this book does not include policies in the document hierarchy, but readers may encounter them throughout the industry governing GCP activities. They may also be found outside Clinical Development to document a company's expectations and requirements in other areas of the corporation, such as Information Technology where they do provide value in setting expectations for employee behavior.

Where are guidance documents?

Documents called "guidance" are not found in the controlled part of the document hierarchy used here, nor in the managed portion. "Guidelines" are also missing. This omission reflects the large amount of confusion about whether documents with these names are binding—do you have to follow them or not? The FDA and EU authorities use *guidance* as just that—something that they recommend, but do not require, be followed. The term *guideline* has similar connotations, but that interpretation is not universal and groups will differ in how they read it. To avoid confusion, companies should avoid using the term or else define it very clearly for their staff. We see in Example 4 that different interpretations can have undesirable outcomes:

> **Example 4**
>
> At a large global company with four office locations conducting clinical trials, three sites interpreted department-managed *guidance* documents as binding and one did not. This created problems when the one site did not follow the procedure in the documents during study startup as written. This resulted in errors in the study database design that required a study database amendment. To avoid future confusion, the company introduced an approach for identifying documents (regardless of title) that must be followed as binding, by using a small lock icon next to the name.

Hard-to-classify documents

This chapter's explanation of the document hierarchy can make it seem that it is always easy to identify the correct document type and to manage it appropriately. In the real world it can be a more challenging as these circumstances show:

- If you have a screenshot in a work instruction, does it become a manual?
- If you do not update your clinical data management plan in six months, do you need to move it into the controlled document system?
- Are checklists a procedural document like a work instruction because they tell you step by step what to do? Or, are they a form because they have check boxes?

In the end, it may not really matter how a document is classified as long as all impacted parties and the controlled document group agrees. The true impact of document type may be how easy it is to find, which can be addressed in a number of different ways as discussed in Chapter 14.

SOP of SOPs

The SOP of SOPs must describe, at a minimum, the controlled document types listed here and how they relate to each other. The methods used to control those document types (see above) provide the rest of the content for the SOP of SOPs. The chapters that follow provide approaches to that control and the example SOP of SOPs found in Appendix 2 puts those approaches into a practical framework.

chapter three

When to have an SOP

SOPs are valuable tools in implementing best practices that ensure quality in conducting clinical trials. But they are not the only tools to introduce consistency, and not every topic warrants an SOP. In this chapter, we explore the question of what kinds of procedures should be documented in an SOP and which types are inappropriate. The discussion in this chapter is strategic and general; Chapter 17 provides more specific practical approaches for companies or groups that are just starting operations and need to develop an entire suite of SOPs from scratch.

Does it belong in an SOP?

The *ICH E6 GCP** definition of SOP, "detailed, written instructions to achieve uniformity of the performance of a specific function," does not provide us with any guidance as to what specific functions or activities require or warrant an SOP. Can it be any function? Does it have to be a regulated activity? The alternative definition introduced in Chapter 1 aims to provide more guidance:

> Written instructions that identify the activities and responsibilities needed to achieve a standard, controlled procedure that *ensures compliance to GCP and applicable regulatory requirements and reflects business needs in support of clinical research.*

The italicized text tells us that SOPs should be for procedures that show compliance with GCP and regulatory requirements, and the definition also recognizes that sometimes a business need may exist that warrants including an activity in an SOP outside of regulatory concerns. A department or multiple departments together may find that they need a standard approach for some other activity that is not strictly required by GCP or other regulatory requirement but is still associated with clinical research.

---

* International Conference on Harmonisation, *Guideline for Good Clinical Practice*, E6.

Is it required by GCP?

So what does it mean that the procedure should show compliance with GCP? A good place to start is with the table of contents of *ICH E6 GCP*. Section 5 lists the responsibilities of a sponsor* and includes topics such as investigator selection, monitoring, and adverse event reporting. Biopharmaceutical companies are the sponsors for clinical investigations and they may delegate some or all of their responsibilities to contract research organizations (CROs) that are then held to the same responsibilities. If the proposed SOP covers activities mentioned in Section 5, then an SOP is probably appropriate. That is not to say that the reverse is true; it would be inappropriate to create an SOP for each of the subsections of Section 5, as some of them do not pertain to procedures and others are too granular. So although there may be one or more SOPs documenting monitoring procedures in alignment with requirements found in Section 5.18, "Monitoring," and perhaps an SOP describing the responsibilities of the study team that supports 5.7, "Allocation of Duties and Functions," it would not be typical to see an SOP called "Multicenter Trials" to align with Section 5.23.

Section 8 of *ICH E6 GCP*, "Essential Documents for the Conduct of a Clinical Trial," is another valuable indicator of the importance of a given procedure in supporting GCP. The essential documents are defined as "those documents that individually and collectively permit evaluation of the conduct of a trial and the quality of the data produced." The tables in Section 8 indicate whether the document is to be held in the files of the investigator or the sponsor, or both. If a procedure involves one of the essential documents filed by the sponsor, then the process followed to create or collect that type of document should be considered an important indicator of compliance to GCP and probably should be covered by an SOP. For example, 8.2.18 lists "Master Randomization List"; this, along with other guidance pertaining to randomization found in Section 5, indicates that randomization is a critical activity that is appropriately governed by an SOP. Of course, each document does not necessarily warrant its own SOP, and those around investigational materials such as 8.2.15 and 8.2.16 may be produced outside the Clinical Development organization, though still filed to show compliance with GCP.

Does it support other regulations?

In addition to *ICH E6 GCP*, sponsors must adhere to the regulatory agency requirements of the countries in which they will file for marketing approval. These requirements, too, should be governed by SOPs.

---

* *Sponsor* is defined in the glossary of *ICH E6 GCP* as "an individual, company, institution, or organization that takes responsibility for the initiation, management, and/or financing of a clinical trial."

For example, the FDA Amendments Act of 2007 requires that parties responsible for certain clinical trials (sponsors) submit the results of some trials to the clinicaltrials.gov databank. It would be appropriate for companies to ensure these requirements are met by creating SOPs for them.

Regulatory agencies also issue guidance documents. Where regulations *must* be complied with, regulatory guidance documents do not need to be adhered to, but many companies try to follow them as they indicate regulatory thinking on a given topic. Guidance documents often provide very useful information about activities that should be governed by SOPs. For example, the FDA's guidance *Computerized Systems Used in Clinical Investigations* (May 2007) has a list of recommended SOPs related to computer systems used for clinical trials in its Appendix A. The European Union publishes rules for medicinal products in EudraLex, Volume 10; Annex IV to *Guidance for the Conduct of Good Clinical Practice Inspections: Sponsor and CRO* (May 2008) lists items to be verified at inspection of a sponsor or CRO site. Section 2.3 has the specific list of procedures (SOP or other) that an inspector reviews to assess compliance with GCP standards and applicable regulations.

Is it a business need?

Even when something is not strictly a regulated activity, a business may decide that a particular process warrants an SOP and most biopharmaceutical controlled-document philosophies support this concept. Business needs can come from within the company or can be perceived as an industry standard that should be adopted. An example of an internal need that is well suited to an SOP is described in Example 1.

Example 1

The process of releasing clinical trial data sets to external parties such as investigators and academic institutions is not a regulatory requirement. Because clinical trial data is both proprietary and confidential, requests for data have to be carefully reviewed, weighing the value of releasing data to the medical community against the risks to the company's interests. An SOP for this process formalizes the steps to review and approve release of the data and ensures that all groups who may receive such a request are aware of the serious nature of inappropriate data release.

The clinical data management plan is an interesting example of a clinical trial document that is not required by regulation or guidance but is considered an industry standard and an auditable document.

Data management plans contain details of computerized systems used for data collection, all the data sources and vendors for the trial, and activities used to ensure data quality and integrity. The information in data management plans could well be documented in other ways, such as with procedural documents and standardized data management files; however, most companies that perform clinical data management require a data management plan for each study and have an SOP that governs its creation and maintenance. Auditors and inspectors will ask for data management plans when reviewing data handling procedures, even though they were not even mentioned in passing in regulatory documents until very recently when they appeared in draft guidance documents. Another practice that is becoming industry standard but is not yet a regulation is the conversion of raw clinical data to SDTM* data sets before analysis. SOPs governing the required mapping and testing of conversion to SDTM are entirely appropriate.

Benefiting both compliance and business

In general, if the topic has significant impact on the company or is an industry standard and it must be uniformly and consistently applied into the future, an SOP is called for. This is especially true when the procedure is cross-functional. When multiple departments or groups have responsibilities in an SOP, the SOP also acts as a service agreement between those departments to commit the resources necessary to perform the work. (See Chapter 10) Although this might not be enough to justify an SOP on its own, it can be a consideration when an SOP is written.

When an SOP is not needed

Companies should never create an SOP that summarizes other SOPs. SOPs are a moving target as procedures, documents, and responsibilities change over time. References to other SOPs are expected and can be helpful (see Chapter 9), but duplicating, rephrasing and repeating, or summarizing what is written elsewhere is asking for problems when the source SOP changes and the two documents diverge. Example 2 is just one case of these surprisingly common, summary-type SOPs.

Example 2

A company created an SOP that listed the key responsibilities of Biometrics staff such as data managers, statistical programmers, and biostatisticians in regards to activities for clinical trials. If this had

---

* SDTM (Study Data Tabulation Model) is a standard structure for clinical trial datasets. While it strongly encourages SDTM, the FDA does not yet require submission of data tabulations in SDTM format.

been used to *guide* creation of all the other SOPs, it might have been helpful. They might have said, for example, that the statistician is ultimately responsible for the statistical analysis plan, clinical study report, and randomization. They could then have used the responsibilities listed to guide the creation of SOPs in each of those three areas. Instead, they reviewed the existing SOPs and then listed in some detail the key responsibilities already found in those other documents. When the other SOPs were changed to reflect changes in process, the responsibilities SOP was not always updated and so did not match or was missing new activities. The Biometrics department eventually retired the responsibilities SOP and replaced it with information in the "About Us" section of the internal web site. One of the first sentences in that section made clear that SOPs, along with department job descriptions, are the ultimate guide for the department activities.

Companies should also not create SOPs for work they do not do! The MHRA[*] writes in Chapter 14 of their *Good Clinical Practice Guide* that they have seen cases of companies that contract out all, or the majority of, trial activity but still write procedures for those activities as if they were carrying out those procedures themselves. The MHRA points out that these "virtual" companies should be writing SOPs that have procedures explaining how they will oversee or assess the quality of the work of the vendor, not SOPs that never apply.

Other SOPs to avoid are those that seem at first to fall into the category of "business need" but that are better addressed by an approach other than an SOP. These SOPs stem from areas that management at a biopharmaceutical company would like to standardize to reduce wasted effort and improve timelines—so they request that an SOP be developed. It is always worth remembering that a company can require employees to do something a specific way, even if it is not described in an SOP. When someone suggests an SOP on emails or file-naming conventions, the goal and intent should be taken seriously but the efforts should be redirected toward training or to department-managed documents. Example 3 shows just such a case:

Example 3

In Clinical Data Management groups, it is not uncommon for the data managers to move from one trial to another as resources are adjusted. One company found that a problem with taking over a trial in progress was an inconsistency in the way study documents were named, and they decided to create an SOP on file naming. A working

---

[*] Medicines and Healthcare Products Regulatory Agency, the UK regulatory authority.

group was convened that spent a great deal of effort discussing what was necessary in the file names. They drafted an SOP that stipulated that the file name was to contain the protocol, type of document (e.g., data management plan, edit checks), version, version date, and other information so that the file name became incredibly long. During the initial review of the proposal, there was such resistance to this convention that the SOP was held up. They only made progress when they decided to reduce the requirements for a file name to match what most people were already doing and to focus their guidance on best practices for versioning. The versioning guidelines were then published as a department-managed document and data managers were instructed to follow it.

Other ways to introduce consistency

As mentioned above, a company or department can require employees to do things a specific way, even if there is no SOP for the activity. Companies do this all the time: they require us to wear ID badges, fill in our time-cards, and go through yearly performance assessments using a specific software application. Of course, we are all told these things are required, and the company provides training as needed. Clinical Development groups can also use this approach: for some activities, setting expectations and providing training gets the same or better result than writing an SOP (as in Example 3). This is particularly true of procedures that are more of a business need than a compliance need. If the activity shows compliance to GCP and other regulations, an SOP is still warranted, but we must remember that SOPs do not take the place of true training. Complex, regulated activities will require training beyond just reading an SOP. (Acknowledging this need for additional training will also help to keep SOPs from becoming too detailed.)

SOP of SOPs

The SOP of SOPs should provide guidance stipulating that SOPs for clinical development should focus on activities that deal with compliance to regulatory requirements or guidelines. But, the SOP of SOPs should also permit SOPs to be created for activities that reflect common industry practices or specific business needs. A manual that provides information for SOP authors can expand on this topic by providing examples similar to the ones found in this chapter.

chapter four

What the SOP should say

Chapter 3 discussed what topics or procedures warrant an SOP; this chapter explains, in a general way, what an SOP should say. We will also touch on a very divisive topic: how much detail to include in an SOP. The discussion of the level of detail to use is related to the question of whether an SOP should combine similar topics into a single document, and that in turn is tied to the question of whether a company can have too many SOPs. Like Chapter 3, this chapter takes a strategic or philosophic view; Chapter 8 provides specific recommendations for SOP templates and structure.

Who, what, when, and where

The first thing any novice writer of SOPs will be told is that the document should explain *who* is performing the work, *what* the work is, and *when* (in what timeframe) it should take place. There is little argument on those, but then the discussion gets complicated. Authors of articles and instructors on SOP writing will then specify that SOPs should *also* include some combination of *how*, *why*, and *where*. Each company will have to decide which of these six options are to be addressed in SOPs, but many companies writing good clinical practice (GCP) SOPs stick to *who*, *what*, and *when* and add *where* as appropriate. This approach is well aligned with the definition for SOP introduced in Chapter 1:

> SOPs are written instructions that identify the activities and responsibilities needed to achieve a standard, controlled procedure that ensures compliance to GCP and applicable regulatory requirements and reflects business needs in support of clinical research.

Who and what

The *who* in an SOP provides the "responsibilities" part of the SOP definition. We provide information to say who is responsible for the activities in question. *Who* is never a named person, it is a person acting in a role. This role is sometimes also a job title, but in other cases the role will be devised for the purposes of the activity described in the SOP. For example, *clinical research associate* may be both a job title and a role in an SOP dealing with monitoring. Similarly, the biostatistician or study statistician may be both a role and job title for an SOP on study unblinding. However, a *treatment assignment gatekeeper* found in that same SOP on study unblinding, who receives treatment codes from outside vendors, is unlikely to be a job title, but instead describes a role a person takes on during study unblinding. In another example, roles of *study leader* and *study manager* were defined for a large company in which the job title of a person leading the study team varied depending on the phase and size of the trial. Some studies had both a study leader and study manager; in other studies the study leader also took on the role of study manager. Commonly used roles, such as study leader, can be defined in a company's controlled document glossary. Others, such as treatment assignment gatekeeper, can be defined in the SOP where they occur. The clear specification of the *who* is important, as can be seen in Example 1:

Example 1

At a medium-sized, fast growing company, SOPs used job titles for Clinical Operations staff. There were clinical trial associates who monitored sites, one of whom would act as a lead site monitor. But some senior clinical trial associates also acted as head of the study team. People in the more senior position of clinical program manager often, but not always, led study teams. Several SOPs were written listing the *who* as the combined clinical program manager/clinical trial associate for study document approvals. Confusion arose around responsibility when approval of the document was requested; was it supposed to be the person leading the study team or the lead monitor? They were both listed, so if the study team had both, was it acceptable to get the clinical trial associate's signature, because that person was far easier to track down than the clinical program manager? Or was it necessary to obtain approval from the clinical program manager, if there was one? This particular company could have benefited from the generic study lead role already mentioned to state clearly when the approval of the study team lead was needed.

The *what* of an SOP is clearly the "identify activities" portion of the definition of an SOP. For each individual step, the *what* describes the action taken or task to be completed. The key word is *action*, and as we see more in Chapter 9 the best SOPs use active verbs for each step undertaken: "distributes the Trial Monitoring Plan for approval," "authorizes the study unblinding," or "ensures that each query is resolved."

Actions may result in an output—this is also part of the *what*. In the GCP world, this is often a document of some kind. The action should create output that is a natural outcome of the work, such as an approved protocol out of the protocol creation process or investigator paperwork that has been filed as part of the study initiation process. An auditor or inspector can clearly check that output as evidence the SOP was followed. Sometimes, it will be necessary to create something that provides evidence that a certain key step was performed. The medical monitor's approval of coding or of severe adverse event reconciliation is one such example. If the coding or reconciliation listings themselves are being retained, the medical monitor could simply sign those, however in large studies the listings can be long and are not always retained, so a form will be created to document that the step was completed. For each activity the procedure authors will need to determine if evidence is needed, and if so, where that evidence is to be retained (see Chapter 5).

When and *where*

The *when*, or timeframe, of the action is often forgotten by SOP writers, but should be an integral part of the action and provided in such a way as to avoid ambiguity. Knowing when to perform a step is critical to performing it correctly. The example actions for *what* in the section "*Who* and *what*" provide more clarity when they read as follows: "distributes the trial monitoring plan for approval *before any sites have begun enrollment*," "authorizes study unbinding *before requesting the treatment codes from the IxRS vendor*," and "ensures that each query is resolved *prior to study database lock*."

Another key component to understanding *when* is to know whether there are any prerequisites. SOP steps are assumed to be sequential. That is, a review step will precede an approval step, and it is assumed that review is complete before the approval takes place. However, this is not always the case as prerequisites may be coming from one of several parallel activities, such as when several documents are drafted in parallel but must be approved in a particular order. A prerequisite can even come from another SOP, as when case report form (CRF) PDFs are produced as part of study database lock procedures in one SOP and those CRF PDFs are then used in the procedures to create the clinical study report, which is governed by a different SOP. SOP templates generally do not have clear ways to indicate prerequisites; some can be included in the wording of an individual process step or action, but some are important enough to warrant a separate line of their own to indicate the importance of stopping and assuring that all required actions have taken place. (See the example SOP template in Appendix 1 for one treatment of prerequisites.)

For most GCP SOPs, the question of *where* does not hold a great deal of significance if taken to mean "physical location," but it is very useful if associated with documents. It adds value to say "Files the approved document *in the trial master file*" or "Files the form *in the study's shared folder* for reference" rather than to say "Files the document" or "Files the form."

How and the level of detail

Now for the difficult question of *how*. *How* goes into the details of a particular action and so carries both value to the organization and risk to compliance. The value in providing the *how* is that it provides the details that employees need to carry out a task and ensures consistency to lead to reliable outcomes. Risk is introduced because specifying the *how* in an SOP means that everyone has to do it just that way and keep doing it that way until the SOP is revised. Details on how to carry out a task are the things that change most quickly in day-to-day work—especially when computer systems are involved. Allowing changes at the detail level can permit an organization to quickly take advantage of more efficient procedures or new features.

Because of the risk of the *how*, many companies have the philosophy that *how* does not go into an SOP, or if it does, it is at the highest possible level—generally specifying only details or tools essential to the quality of the outcome. For an SOP that avoids *how*, a task might read "draft the study protocol." Going to the next level of *how* to include an essential tool, it might read "draft the study protocol using the most current Therapeutic Area template." Introducing yet more detail takes the action from one sentence to many steps, as we can see from just the beginning of a detailed version of the step we started with:

- Copy the Therapeutic Area template from the XYZ Document Management System to the study's protocol folder on the shared drive.
- Replace the placeholders for the study identifier and study name throughout the document.
- In consultation with the study team, fill in the Visit/Procedure Matrix.

Many other steps would follow because detail often begets more detail so that the moment the SOP says what part of the protocol template to customize for the study first, the question then becomes what part to customize next, and so forth.

Over the years, the level of detail in biopharmaceutical SOPs has swung back and forth between extremes of detail. A company may start with a high level of detail as shown in the bullets above, and this detail will be appreciated by the staff as it provides useful guidance on the activity. Later, they find that they are stuck having to update the SOP when the document system changes or when groups find that filling out the visit or procedure matrix first does not work as well as it used to when they were doing mostly Phase I studies. The company then changes the strategy and requires that SOPs be revised to go to a minimal level of detail, and then they find that staff no longer creates protocols of the same level of quality. Rather than find a happy medium, the pendulum continues to swing from one extreme to the other.

The best solution is, not surprisingly, likely to be somewhere near the medium level of detail that is so often passed over. SOPs should identify tasks at a high level then go the next step and list any *required* tools, hand-offs, or approaches that are known to improve the quality of the outcome or efficiency of the process. Any additional details that staff members need to do the work can be put into supporting documents or training as described in Chapter 2. Those documents are easier to update and have more flexible document templates that can accommodate useful screen-shots and examples.

When setting an SOP strategy regarding the level of detail, it is a very useful exercise to repeat the mantra "an SOP does not replace training." Although it is true that everyone who carries out the procedures covered in a particular SOP must be trained on that SOP, it does not follow that training on the SOP alone teaches people enough to enable them to carry out their jobs. This mantra will also prove itself very useful to SOP writers when they find themselves getting bogged down adding details to an SOP, because those details are not available anywhere else and are "really important." It is much more effective to keep those important details out of the SOP and create new supporting documents, which can be revised much more quickly than an SOP. It is more effective still to create or revise training to align with the SOP and to refer to those documents that contain the details that support the SOP in the training materials.

Cover one topic or many topics?

In Chapter 3, we addressed which topics should be governed by an SOP. We have not yet addressed how many of those topics should be in a single SOP. Some companies will write an SOP with very broad coverage of a process or for a process such as "Study Initiation," "Data Handling," or "Activities of Development Drug Safety." Putting all closely related topics together in a single SOP ensures that everyone understands the end-to-end process and also avoids possible gaps, inconsistencies, and even conflicts by providing the entire process. Although this is an admirable intent, the length and complexity of SOPs with broad coverage of multiple topics introduces a complexity in actual use (as shown below in Example 2) that makes it hard to justify. It would be better to break the topics into several SOPs and perform a gap analysis to ensure nothing was lost in separation.

Example 2

A company had a single SOP for their Drug Safety group; its purpose was to define the processes and interactions with other departments to ensure compliance with applicable relations. The SOP was thirty-two pages long. The summary of responsibilities covering all groups in Clinical Development and many subgroups required four and a half pages. Eighteen pages described eleven different activities. Because all departments were mentioned in this SOP, nearly everyone in Clinical Development had to read it all even if much of it would not have applied to any given person.

Too many or too few SOPs

The question of how many topics to cover in a single SOP is linked to a question that larger companies face after many years of development and several mergers or acquisitions: do we have too many SOPs? Over time, these companies will have ended up with a mix of SOPs from various points in their history that have valid scopes but present a confusing library to users who are also not sure whether legacy SOPs or new SOPs apply to any given study. Management will typically recognize that there are too many SOPs and undertake an initiative to correct the situation. One such case is described in Example 3, which has a company move from having too many SOPs to having too few SOPs:

Example 3

After a merger, one large global company recognized that their hodge-podge of SOPs would not be good for the organization going forward. The Controlled Document group had a good plan (with management support) to have representatives from all departments in Clinical Development and all locations meet for a few days and identify topics for critical SOPs. The lists of SOPs that emerged from those meetings were a reasonable set that covered the clinical development process from end to end. Those SOPs would then be written afresh to represent the ideals of the newly merged organization. Teams were given a year to create and put into effect all SOPs, but after two years, fewer than half were done. The company tried again by identifying a small handful of SOPs, fewer than in the first round, that would be applied globally—this initiative went forward, but all the departments wondered how just a few SOPs could provide enough information for them to do their work. The departments concluded that their only option would be to have more department-managed documents and began to increase resources in that direction.

Both of these situations, too many and too few SOPs, will lead to compliance problems. In the case of too many SOPs, auditors can readily tell that it is difficult to know which SOP applies in a specific circumstance. In the case of too few SOPs, the details that people need to do their work are all pushed down to supporting documents such as manuals, work instructions, and department-managed documents. If the details are all in supporting documents that are controlled, you still end up with a larger number of controlled documents; they are not SOPs but they are still auditable. It may be that these documents are easier to maintain than SOPs, but you may lose the ability to document cross-functional activities because manuals and work instructions are frequently limited to a single department. If the groups depend too much on department-managed documents, again the total number of documents is greater than just the number of SOPs, and in this case it is possible that the elements of document control outlined in Chapter 2 are not being well applied at the department level—a situation that can lead to audit findings also.

Large companies will have need for a larger number of documents to produce consistent work globally or even within one large department. Those documents may be SOPs, other controlled documents, or department-managed documents. Unfortunately, there is no solution to reducing the total number of documents of all types, but clear document hierarchies and attention to ensuring no overlap and no gaps in the development process should keep potential compliance issues to a minimum.

SOP of SOPs

The SOP of SOPs may not be the best place for guidance for SOP authors on what an SOP should say, but a manual would work well. Guidance for authors should include these key messages:

- An SOP should cover the *who*, *what*, and *when* of a procedure but only provide instructions, *how*, when that entails tools or steps essential to the quality of the result.
- SOPs do not replace training, so details that someone would need to carry out a task should be included in training and/or in other documents such as work instructions and manuals.

To allow flexibility in both training and revision, SOPs should cover only closely related GCP topics in a single document. However, care must be taken to avoid making SOPs so granular that the set becomes large and gaps and inconsistencies are introduced.

chapter five

Where to put the output

In Chapter 4, we saw that the general philosophy of writing SOPs includes addressing the question of *what* action to take and identifying any output generated. In good clinical practice (GCP) SOPs, the *what* that is the output of a procedure very often includes or even focuses on a document or form. Many writers of SOPs are not aware that they need to say not only *what* is created but what to do with that document or form that is generated as part of the process (*where* to put the output). Some more experienced writers may know to say something like "retain in the study files," but that is not clear enough. Does it mean storing it in the shared folders for the study until the study is over? Does it mean printing it out and putting it in a folder? Should it be submitted for offsite storage? Should it go into the study's trial master file (TMF) to be available for inspection? Because many people who work with SOPs are not necessarily familiar with practices and trends regarding documents and the TMF, this chapter will start by providing some background, then make the connection back to SOPs.

What is a trial master file?

The trial master file, or TMF, is the name given to the set of documents related to the conduct of a clinical trial that must be retained to meet regulatory requirements. Section 8 of the *ICH E6 GCP** is titled "Essential documents for the conduct of a clinical trial" and says the following in Section 8.1, "Introduction":

> Trial master files should be established at the beginning of the trial .... A final close-out of a trial can only be done when the monitor has reviewed both investigator/institution and sponsor files and confirmed that all necessary documents are in the appropriate files. Any or all of the documents addressed in this guidance may be subject to, and should be available for, audit by the sponsor's auditor and inspection by the regulatory authority(ies).

* International Conference on Harmonisation, *Guideline for Good Clinical Practice*, E6.

The contents of the TMF maintained by a biopharmaceutical company must contain, *at a minimum*, the documents listed in this section of *ICH E6 GCP* as belonging in the files of the sponsor. It is most common for TMFs to contain (many) additional documents judged by each company as being important to "demonstrate the compliance of the investigator, sponsor and monitor with the standards of Good Clinical Practice and with all applicable regulatory requirements," as the introduction to Section 8 sets forth. Taken together, the documents in the TMF permit the evaluation of the conduct of the clinical trial and the quality of the data produced.

In December 2012, the European Medicines Agency (EMA) published a draft guidance titled *Reflection paper on GCP compliance in relation to trial master files (paper and/or electronic) for management, audit and inspection of clinical trials*. This publication provides very helpful background information to understand the scope of the term *trial master file* including these key concepts:

- Documents in the TMF come from many sources: "In large organisations, the TMF could include documents from across a variety of different departments and systems other than clinical operations, for example, Data Management, Statistics, Pharmacovigilance, Clinical Trial Supplies, Pharmacy, Legal, Regulatory Affairs etc., as well as those provided or held by CROs*" (Section 4.3).
- It is not necessarily a single "file cabinet" (electronic or paper): "Sometimes documents may need to be located in a separate location to the main TMF records, for example those that contain information that could unblind the study team" (Section 4.3).
- Key documents from following SOPs must be retained to demonstrate compliance: "Any quality record produced from following a quality system procedure must be retained in the TMF to demonstrate compliance .... Examples include evidence of QC checks, documentation on Regulatory Green Light, Database Lock Forms etc." (Section 5.4).

Later, in Section 9 of the reflection paper, the writers list common problems with TMFs from inspections and they specifically mention incomplete TMFs: "In some cases resulting in additional inspection days required. This is often as a result of the contents being restricted to the contents of ... Section 8 documents."

In summary, the TMF is a collection of documents generated during conduct of a clinical trial and collected to demonstrate compliance to GCP, the study protocol, and SOPs. The "file" itself is a logical collection, rather than a physical one, as the documents may be stored in multiple locations

---

* Contract research organizations.

and can be electronic and/or paper-based. The TMF is one of the most important focuses during regulatory agency inspections.

Contents of a TMF

Although the Section 8 essential documents give us a start and the EMA reflection paper provides guidance, there is no other complete or definitive list of items to include in a TMF available from regulatory agencies. Each company has had to create its own list based on its own SOPs and computer systems, and in reality this has meant that individual departments in each company have created lists. Many companies had, and perhaps still have, a Clinical Operations TMF, a Clinical Data Management TMF, and other "study files." Many of the departments will define a list of items they feel should be in a TMF and then create forms to be used to check it. The departments are left making the assessment of what should and should not be included and how it should be organized.

In 2008, a group within the Drug Information Association (DIA) formed to address the strong industry interest in a standard starting point for the contents of a TMF that would include the essential documents from *ICH E6 GCP* and other documents that were generally recognized as important by the industry. They specifically included documents from groups beyond Clinical Operations, like Data Management and Biostatistics, to define a single complete TMF structure. In 2012, this group released the first version of a document known as the "TMF Reference Model."*

Overview of the DIA TMF Reference Model

To keep the description of the model at a very high overview, the Reference Model is essentially a large table with rows of types of documents (known as "artifacts" in the model). Examples of artifacts are:

- Monitoring Plan
- Informed Consent Form
- Monitoring Visit Report
- Serious Adverse Event (SAE) Report
- Data Management Plan
- Randomization Plan
- Completed Case Report Forms

These are grouped into "zones" such as "Trial Management," "Regulatory," "Statistics," and "IP[†] and Trial Supplies." There are further

---

* This document is available free of charge at the DIA website, www.DIAHome.org
[†] Investigational product.

groupings of document types and many other columns used to provide information on what the documents are, where they come from, and where they are stored. Most of the document types, or artifacts, can be thought of as folders; there will be multiple versions of the Monitoring Plan in a TMF for a Phase III study and multiple Monitoring Visit Reports for multiple sites for most studies. There will be multiple SAE reports for most studies and they probably are stored in the company's validated SAE reporting system.

Company TMF reference models

Some companies will still develop their own TMF structures and lists of contents and others will begin with the DIA TMF Reference Model and customize it. The process of creating a company-specific model is not dissimilar to customizing the DIA model, so this section will focus on how companies adapt that model to their needs. To use the DIA Reference Model, each company starts with the most recently released version (to ensure it reflects recent updates to regulations) and then modifies it as appropriate for the company. They list the actual names of documents they use; for example, in the "Data Management Plan" artifact, a company may file its "Data Handling Plan"—because that is what they call the document that is a data management plan for the study. Companies will also remove rows for documents that they never use. For example, the randomization plan may always be included in some other document, perhaps as part of the Interactive Response Technology (IRT) requirements document, which has its own artifact. Companies will have to be sure to add to the reference model any documents required by their own SOPs that do not naturally fall into any of the given artifacts.

Large companies may have more than one reference model. They may have one model for the early research group conducting mostly Phase I studies, another for the later stage unit, and yet another for the Medical Affairs, Phase IV group that only outsources studies. All of these will be valid TMFs with the required documents, but the specifics of what is never used, what the document is actually called at that company unit, and who is responsible for producing it could vary enough to warrant multiple models. These units may also be working to different SOPs, again impacting what documents will be generated. A TMF governance group, with representation from all impacted departments, will release versions of the company TMF models and review requests for changes.

Paper TMFs and electronic TMFs (eTMFs)

The structure defined by a reference model can be applied to paper TMFs or it can be built into an eTMF system. When the model governs paper TMFs, it can be more flexible and the folders and their contents are not restricted. If a new SOP requires a new type of document, studies already started can add it to an existing folder if appropriate or simply create a new folder. When a reference model is used for an eTMF, software implements the folders and may restrict the documents accepted into a given folder. Because the software will have had to have been validated, it will be under change control, which will significantly impact the kind of changes that can be easily made. Whereas the rule in paper systems is straightforward and left to users ("we will add it to the file if we have the document"), the rules in eTMFs are implemented by the software and must be clearly laid out at the time the change is implemented. Changes that are not limited to new studies but are partly retroactive to existing studies are particularly challenging. In Example 1, we see an even more complex case.

The TMF connection to SOPs

The TMF and SOPs are tightly connected and the connection is made even more apparent as the DIA's TMF Reference Model is adopted and companies begin to implement eTMF systems. If an SOP says to file a document in the TMF, there has to be a clearly defined location for it in the TMF. Conversely, if the TMF model calls for a given document but the SOP is revised or retired so that document is no longer created, the TMF model must be revised so that document is not required in the future. It may even be the case that the *names* of documents created as part of a procedure will have an impact on the TMF (see Example 1).

Example 1

A large company had been using a vendor electronic data capture system for several years. In implementing the system, they designed a set of study specification documents that covered the electronic case report form specification, edit checks, study configuration parameters, and visit structure. After having gathered a great deal of experience, the company decided that the process of study setup could be simplified, and management initiated a process optimization project. The outcome of the project was a different set of study specification documents—with different names and different content spread across two artifacts (database design and edit check specifications). In the middle of this process optimization, the company separately began implementing a new eTMF system based on the DIA TMF Reference Model. The global eTMF implementation team opted for enhanced reporting and so implemented a drop-down list of documents that could be accepted for each artifact. Because existing studies would continue to use the pre-optimization list of documents and new studies to use the new list of documents, the system had to support *both* variations in the impacted artifacts. Adding yet another difficulty, in the case of one of the document types, studies would transition from the old documents to the new and so could have both in the same TMF.

The TMF reports the implementation team wanted to support depended on keeping the document count accurate; so depending on the study the number of documents expected in the two artifacts would be different. Keeping the counts accurate in this complex situation in order to provide reports did not seem to warrant the effort required. Some members of the implementation team argued for a more flexible filing arrangement where it would be the responsibility of the person filing to ensure the document was placed in the right artifact and identified by an appropriate name—no drop-down lists and no pre-set expected counts, as it would be in a paper TMF.

Not all documents belong in the TMF

Just because a document is generated during the course of a study does not mean that it is destined to be part of the TMF—and therein lies the art of document management. When considering whether or not to retain a document or form, we first have to determine if it is useful in showing GCP compliance, adherence to regulation, or quality of the data. If it is, then it is a candidate for filing in the TMF. But we should also further ask if that document or form actually results in the information regarding compliance being recorded *elsewhere* in a key study document or other system. If there is other evidence, we do not necessarily have to retain the document being assessed. Several examples will quickly show how difficult the decisions can become:

- Many of the items that would be categorized under *correspondence* in a TMF—usually emails, these days—would not need to be retained if the outcome of the correspondence is adequately documented elsewhere. If the medical monitor notices a data issue and emails the data manager to include an additional edit check, the email itself does not have to be stored because the addition of the edit check will be done in a controlled way, with updates to a specification document that has a version history. However, when email is the *only* record, as would be the case when email approval for a change to a study document is permitted, then those emails should be retained.
- When logs or "trackers" are used as communication tools, there is a question of whether or not to retain the log. It is typical for study team members to use tables or logs to communicate issues. For example, issue logs are often used to communicate items of concern in the data to a data manager, who will then review and issue a query to

the site, if appropriate, or respond to the originator that other queries already address that concern. The evidence is the query to the site along with case report form (CRF) data (and audit trails for electronic data capture). The issue log does not have to be retained after the study lock unless it documents irresolvable issues that do not appear elsewhere. Similar logs are used in communication with coding groups and with external vendors providing non-CRF data in electronic format (such as lab data). Trackers are often used to track study progress to clean subjects and, again, do not add to what is already recorded in the data records and associated resolved queries.

- When checklists are used as a work-aid to assist in performing a complex task and they are not the only evidence of the work being done, then they do not have to be filed in the TMF. If, however, they are the only evidence of certain key activities being performed, then they do need to be retained. Study-build checklists are a good example here. Study-build checklists have steps that are documented elsewhere, for example, obtaining approval for the CRF from the study team on a controlled form. But there may be other crucial steps for which there is no other evidence if a form is not available to document them. Second reviews of the build by an independent person, a common quality practice, would not necessarily have separate output or evidence. To provide evidence that the review took place, a company can (1) create a form for the reviewer to sign, (2) sign and retain the checklist, or (3) decide that it is not worth documenting that step because missing the review step is low risk to the quality of the data.

In all of these cases, any SOP governing the activities should clearly provide the document disposition. As we will see in Chapter 9, a section of the SOP template that specifically addresses document disposition helps remind SOP authors to go through these discussions before the procedure is rolled out, rather than at the end of a study when staff are faced with the question.

Whether or not to retain certain documents is an area where reasonable and experienced people will differ in their opinion, and this may also include representatives of regulatory agencies! The most important thing is to have the discussion of whether or not to retain a given document internally with the understanding that some of those people consulted or in a decision making position will lean toward retaining more documents and others will lean to retaining fewer. Whether the decision is to file a document in the TMF or not, the decision must be explicitly noted in SOPs, supporting documents, and, as needed, in the TMF reference model.

SOP of SOPs

Every SOP that creates a document or makes use of a form or checklist should say explicitly what is to be done with it. Enforcing this requirement is usually accomplished through the SOP template, rather than in the SOP of SOPs. Even with the requirement spelled out in the SOP template, SOP authors from the departments in Clinical Development cannot be expected to have an intimate understanding of the corporate TMF model being used, nor of the filing options open to them. A review of each SOP by someone familiar with these document management concepts in general and the TMF specifically will be of high value to the organization. This review can be performed during compliance review of SOPs or as a separate review by the TMF support or governance group as demonstrated in the example SOP of SOPs in Appendix 2.

section two

Writing, reviewing, approving, and posting

chapter six

Who writes SOPs?

In the discussions of the foundations of developing and writing SOPs covered in Section 1, the idea of the SOP author being guided by the SOP of SOPs and the SOP template was introduced. Who is this author of the SOPs in an organization? The original writer should be someone from within a department function, but this person should be assisted and supported by staff from the group responsible for cross-functional oversight of SOPs and the controlled document system—which we have been referring to as the Controlled Document group. This chapter will identify qualities that identify good candidates for authors from a functional group and touch on how the Controlled Document group can support their efforts.

Which department provides the author?

In the simplest case, an SOP will only include actions taken by roles from within a single department or functional group, in which case the principal SOP author will clearly be a member of that group. Most SOPs, however, will include several roles from different departments that have responsibilities for one or more tasks. In this case, there is usually a department or group that sponsors the SOP by providing the writer—usually this is the functional group that is at the center of the action or that is primarily responsible for the key outcome or output of the procedure. Such a group is often known as the *business process owner* (BPO) of an SOP.

When a process impacts several functions equally, the author may be identified from the impacted functional group that has appropriate resources (people) available. Even in this case, a single function should take on the role of BPO, because in addition to providing the author the BPO representative also acts as the point of contact for adjudicating feedback from the review of the document, performing initial training assessment, and reviewing change requests filed for the document. Although these may appear to be mostly administrative tasks, they do provide the opportunity for the functional group identified as BPO to have some amount of extra control over the wording in the final document and over future revision of the document. The example SOP of SOPs found in Appendix 2 incorporates the concept of BPO.

Selecting an author

A guiding rule of SOP development across industries is that people who understand the process should write the SOP. It cannot be stressed enough that the person documenting the process must understand the details of process as it will be performed at that particular company. This means that SOP writing will typically fall to more senior or experienced staff. This does not, however, mean that managers will write the SOP. Managers should only be called upon to write SOPs if they are very familiar with the details of the process. This usually (but not always) means managers at smaller companies. At companies where managers mostly oversee staff and project resourcing and are not closely familiar with the critical aspects of the work being performed, a manager would not be a good candidate for writing the SOP.

Where the task is the revision of an SOP for a process already established at the company, the best writer is someone who has performed steps in that process. In the case of a new process, no one may have experience with that particular procedure at the company, but those staff members who have participated in the working group that drafted the process, discussed it at length, and perhaps participated in a pilot would be the best ones to draft the SOP (see also Chapter 7).

In addition to experience, logical thinking and clear writing are also desirable qualities in an SOP writer, but these are harder to identify because people are hired for the work they can do, not the procedures they can write. Managers can still identify group members who have shown an ability to clearly describe a process, perhaps in a presentation or informal training. It is also true that people who *want* to try writing an SOP often have the right mindset to do so well—so asking for interested candidates is also a good approach. It is not uncommon for groups to assign SOPs on a rotating basis to staff members to both spread out the effort and also provide development opportunities for less experienced staff. Unfortunately, assigning SOP writing and revisions to staff as part of yearly goals without their input and willingness will often lead to unhappy people writing poorly specified SOPs. Not every experienced, detail-oriented staff member will be able to write quality SOPs.

When a candidate SOP author has been identified and has agreed to the project, the time needed to do the work must be factored into that person's overall workload. One way that functional groups fall behind planned timelines in creating or modifying SOPs is to assign SOP writing or revisions and then fail to explicitly allocate time in the author's schedule and work assignments. In the end, the "real work" will always be given priority and the procedure writing will lag behind if it is not actively planned for.

Because of the problem of finding capable staff members who have the time and talent to write and revise SOPs, larger Clinical Operations and Clinical Data Management groups have begun creating infrastructure subgroups within their organizations whose remit includes writing and managing process documents. (Often overseeing and providing training is also included.) These subgroups generally recruit staff members who have shown an interest in and facility with process documents, document management, and compliance questions. When people transfer in to these infrastructure groups, they no longer do study work. This situation has the powerful advantage of identifying people whose job it is to deliver and manage documents. Because they no longer work actively on studies, they may not come up with new processes themselves or identify needed process changes; however, they are in a very good position to fine-tune documents written by others, check for consistency with other controlled documents, oversee review, and otherwise ensure documents are released and properly imbedded in the organization. This kind of department process-group can also act as the liaison to or point of contact for the Controlled Document group to shepherd SOPs through the revision, review, and approval process.

Clinical Development's Controlled Document group

Even the smallest company needs SOPs, so even the smallest company should identify a person or create a small group whose responsibility is to manage controlled documents. This group would typically fall under the Regulatory or Quality Assurance groups. Because good clinical practice SOPs and the controlled document hierarchy are different than those used in good manufacturing practice (GMP), the best arrangement has this group reporting within the Clinical Development hierarchy, though they may share software systems for the management of controlled documents with other branches of the organization.

The people working in the Controlled Document group are not the authors of the SOPs for the Clinical Development functions—with the exception of the SOP of SOPs and all closely related SOPs or supporting documents governing controlled documents. This group's job is to assist the authors of SOPs in producing quality SOPs that are in compliance with regulations and consistent with other Clinical Development SOPs. The level of this support may vary from company to company, with some groups providing a great deal of assistance in editing, formatting, and routing for review, and with other groups providing only minimal support in the form of compliance review and posting. In all cases, the Controlled Document group would be responsible for the administration of SOPs, which would include moving SOPs into the controlled document system, providing editable copies for revisions, managing retirement, providing copies during inspections, and so forth.

SOP of SOPs

To leave the greatest flexibility in adjusting to business needs and realities, the SOP of SOPs should only touch on the question of authors of SOPs to the extent of saying that the writer *should* be someone with knowledge of and (when possible) experience in the procedure being documented. In most cases, the writer should be drawn from the functional group that is most responsible for the procedure (the BPO). The SOP of SOPs or a related SOP on the management of SOPs will have a great deal to say on the responsibilities of the Clinical Development Controlled Document group.

chapter seven

Document a stable process

One of the biggest sources of noncompliance to SOPs is writing and posting SOPs for a process that is untried or not stable. SOPs should *never* reflect a theoretical or untried procedure. There are ways to test a procedure taking into account any SOPs that are already effective, and there are ways to document a process when no SOP governs it. This chapter provides examples of what can happen when untried procedures are put into SOPs; it proposes approaches to creating processes and the related SOPs that can be followed on the day they are posted as effective.

SOPs should not be theoretical

When an SOP is posted as effective, it must be followed by all studies to which it applies, or a deviation must be filed. If the process documented in the SOP has not been thoroughly tested, the SOP may be posted with steps that cannot be followed as written, it may be missing critical steps, or it may include inefficient or burdensome activities. This happens at companies of all sizes all the time, as shown by Examples 1 and 2. At small companies, SOPs written from unstable processes occur as staff rush to create new SOPs for activities that may not have been performed at their organization more than a few times. At many companies, introducing new software systems—something that can happen multiple times over a short span of years—causes this problem as the implementation team revises existing SOPs without really understanding the impact of the new software. During mergers and acquisitions, this issue comes up again as the parent company decides to revise SOPs to reflect best practices taken from the two firms.

Example 1

A growing company that was revising its clinical data management SOPs, which had been written when it was small with a limited staff, assigned a senior manager from the group to author the revision. When a few experienced data managers saw his draft, they uniformly said, "That is not the way we do it!" to which he replied, "That is the way you *should* be doing it." The data managers were not sure the task *could* be carried out as recommended by the manager—certainly no one had even tried. This issue went around and around and the SOP was held up for months, but in the end the data managers prevailed.

Example 2

A company moved to conducting all of its studies with electronic data capture (EDC) using a specific, vendor-supplied system. They began starting studies in the new system using SOPs that had been developed based on extensive piloting (see below) and were adequate. They also developed a study closeout SOP, as their SOP for closing paper-based studies could not be applied. Although they used senior and very knowledgeable staff from several functional groups to develop the new closeout procedure, they ended up posting an SOP for EDC closeout that was theoretical because they finalized it before it had been used for *any* studies. From the very first study that followed it, they realized that they could not perform study database lock with the steps as written. After similar experiences with additional studies, they had to retire the closeout SOP without a replacement and work from study-specific, detailed checklists until the process could be ironed out and a new SOP drafted based on actual studies closed.

What the groups in both of these situations did not understand was that it is better to have no SOP (and document process on a study level) than have an SOP you are not following.

Testing a procedure

Nearly every new procedure should be tested before an SOP reflecting that procedure is posted and must be followed. The best way to try out a procedure is through the use of "pilot," or test, studies. In an ideal world, companies would pilot procedures on sample studies, but in the real world, these pilots are going to be conducted on actual studies with real subjects and data that may end up in a submission to a regulatory authority, so all care must be taken to minimize risk, assess compliance, and document for future reference or inspection *exactly* what was done. Because of the use of actual studies, before any pilot can be conducted, the following must occur:

- The process to be followed must be thoroughly vetted by knowledgeable staff.
- An assessment must be made of whether any SOP deviations are required.
- Studies that represent an appropriate range of practice must be selected to conduct the pilot.
- Documentation about what process was followed must be provided for the studies involved in the pilot.

Draft and vet the process

Although it may seem counter to the idea of testing a procedure, the team working on the procedure must first draft the process to a good level of detail in order to carry out a valid pilot. For new procedures, this would typically be a detailed process map, though it may take the form of an SOP draft, table of steps, or even detailed training slides. For revisions or reengineering of a process, edited versions of existing materials may suit the need. The team must then create solid working versions of any templates or forms that the procedure calls for. Because translating a workflow into text often identifies areas where steps are vague, working from a draft process document can be more valuable than working from a process diagram, if time permits.

When a complex process has been developed, there are risks to the earliest pilot studies. Simulating the process as much as possible before using it on a study can mitigate the risk. A process walk-through can be a powerful tool when done right. In the best process walk-throughs, experienced staff members representing each of the functional groups involved in the activities will commit a significant amount of time to evaluate the new procedure. This could easily range from a half day to several days; but it is certainly more than a one-hour meeting.

The pilot team leads the walk-through and starts by describing the new process to the functional group representatives. The pilot team may use the slide presentation that will be used later to describe the new process to study teams that are candidates for pilots, and they also provide the process diagram (see Chapter 8) and/or draft SOP. The functional group representatives act out each step. They may pass around pieces of paper representing documents, files, or other materials as in Example 3.

Example 3

For a revision to an SOP that governed the process for building an electronic case report form (eCRF) at a large company, the walk-through involved staff representing the roles from Clinical Data Management, EDC Programming, External Data Acquisition, and the Data Standards group. The representatives used a pretend eCRF specification. The person in Data Management who was expected to create the initial draft of the eCRF specification started with the paper in hand. The group then handed the paper around as the pilot group explained the process. At each stop, the person holding the paper explained what they expected to do to that specification as part of the real process. This exercise quickly pointed out that there was too much handing of the specification back and forth during the steps to check eCRF elements against corporate data standards.

The experience from the walk-through resulted in a revision to the proposed process to address the extra handoff prior to a pilot being conducted.

Someone from the pilot team should be taking notes during the walk-through regarding items to be researched or changes to be made to the workflow. The functional group representatives should review the revised documents to ensure that they accurately reflect solutions to issues uncovered in the walk-through.

Assess the need for deviations

The draft procedure with its associated templates and forms will help the pilot team assess whether or not it will be necessary to file prospective deviations to existing SOPs. This is a critical step that cannot be avoided. If there is already an SOP governing the process in question, there is a very good chance that any change to the procedure will violate the existing SOP. Most companies will have a process for declaring a prospective deviation (see Chapter 12). The group requesting the deviation will have to provide the information for what is being followed instead of the SOP (e.g., the draft process), what the impact on study quality will be (presumably the same or an improvement), and they may have to list all of the studies to which this deviation will apply. At some companies the deviation may require an explanation of how all impacted staff will be made aware of which process to follow.

The approach to selecting studies is described below, but in cases of major process reengineering the pilot may have to be more open ended and add studies over time. In this case, the named list of studies is not known at the time the deviation is submitted. It may be possible to work with the Controlled Document group to file the initial deviation and later provide the complete list of studies as an addendum to the deviation as in Example 4. At other companies, it may be necessary to file a deviation for each study separately as they are added to the pilot.

Example 4

A company planned to streamline the user acceptance testing procedure for their EDC studies. The process optimization group came up with a procedure and prepared the documentation. They identified the kinds of studies that should be involved in the pilot based on geographic site, phase, and therapeutic area as recommended later in this chapter. Because they needed a specific range of studies, they were going to have wait for the right kinds of studies across development projects to start up, and because study starts can be delayed for all kinds of reasons they would not know the true, complete list at the start of the pilots. They were able to get agreement from their good clinical practice compliance committee to submit the deviation and then provide the studies to be added to the deviation as they became known.

When a process is new or covers a new activity, it may not conflict with an existing SOP and so a deviation will not need to be filed. It is nearly always acceptable to do more than is listed in existing SOPs, as in the case of a company that added a level of testing during eCRF build. In addition to the unit testing by the programmer and the user acceptance testing by the study team, they added a very detailed "quality testing" to address a problem with the quality of studies released to production. They only added to the existing SOP so did not have to file a deviation. They did, however, update the SOP after the pilots to reflect the new practice.

Select studies

When a good draft of the procedure is ready for piloting, studies to try it out on have to be selected. All but the smallest companies should strive to identify more than one study—perhaps one Phase I study since they start and end quickly and also a later phase study as these will typically be more complex. At larger companies, studies from different therapeutic areas should be included, as the staff that support these studies may do things differently. At companies that have multiple corporate sites, studies being managed from each location should be represented; this is particularly important for multinational companies where offices in different countries may have significant differences in the way they run trials. In the cases of mergers or acquisitions, again select studies managed by the different, original, organizations.

When candidate studies have been identified, it is essential to get the agreement of the study teams. Teams with very tight timelines may not be open to being the first to try out a new, untested process. On the other hand, if new efficiencies are being introduced, those same teams may be very enthusiastic about taking advantage of them on the chance of making it easier to meet the timelines. Often, the group drafting the procedure will prepare a slide presentation to be given to the study teams so that they are very much aware of what changes or new procedures are being introduced and what the likely impact on, or risk to, a study could be.

Document the process used

A pilot is most likely to take place on an actual clinical trial and, because of the reality of study and project timelines and availability of resources, most pilots only follow the new procedure rather than running both procedures in parallel. Because of this, it is essential to document which process was *actually* used, to show that it was a controlled process (even if it may not be fully debugged) and that the quality and integrity of the trial data were not impacted. When a deviation is filed, the information about the process to be followed will be submitted with the deviation request, but it is also best practice to document that a deviation is being followed directly with this study. This is also true of a fully new process for which a deviation was not required.

Because deviations and associated documents are often not easily located in controlled document systems and may not be linked to the individual study, it should be possible to find the information with the study. Ideally, the information would appear explicitly in a document already associated with the study and the activity in question. For example, a deviation from a monitoring SOP for purposes of piloting a new, reduced site monitoring process could be noted at the very beginning of the trial's monitoring plan. A change in clinical data management procedures would be documented in the data management plan, and so forth. In some cases, it may be necessary to document that a new or revised process is being followed in a study note to file, though this is not the ideal method (see Chapter 12). In addition to noting the deviation in normal study documents, any materials for the pilot, such as the process diagram, draft SOP, and even training materials, should be accommodated in the trial master file to ensure that information is available in case of inspection. Some companies may also require that training records for the new process be formally filed either in the training system or in the trial master file.

When you cannot pilot

There are some cases when a pilot is not possible. This is certainly true for small companies just starting trials, but there are also some compliance groups who will not permit pilots on actual studies, maintaining that the risk to an actual study is too great to permit an interim process. (One does wonder then about the risk to all studies when an untried SOP is posted.) And, it is also logistically difficult to pilot on new software systems or new releases because studies cannot be run on software that has not been validated—and validation is all too often the last step before widespread implementation. But all of these situations can be addressed or ameliorated.

Chapter 17 discusses in depth approaches for small companies creating suites of SOPs. In the case where pilots are not permitted, the implementation team must use all of the techniques at their disposal, especially process walk-throughs. In the case of completely new software, companies should plan on a staged rollout where the old system is in place alongside the new system, allowing some studies to pilot on the new system before the other studies cut over. Unfortunately, this is usually not possible for version upgrades, because only one instance of the software is likely to be available at any given time; in this case, we have to fall back to the same approach as when pilots are not permitted and vet the process as thoroughly as possible before it is introduced as a package with the new software.

SOP of SOPs

It is not possible to go into details regarding the requirements for a stable process in an SOP of SOPs. The best that can be done is to indicate that SOPs are to document only stable, controlled processes. It is very likely that the significance of a sentence to that effect will be too subtle to get the attention it deserves. This is where the Controlled Document group can provide support. When they receive a new document for processing, they can simply make inquiries regarding how fully the revised or new procedure in the SOP was tested, and they can provide guidance as to the importance of posting only SOPs that reflect stable processes that can actually be followed.

chapter eight

Mapping a new process

We learned in Chapter 6 that the most appropriate person to write an SOP is someone who is experienced in the procedure to be documented and that, ideally, that person has a logical mindset and can present ideas clearly. People like this often do not have previous experience writing documents like SOPs. This chapter provides an introduction to how SOP writers can take a process and translate it into a workflow and then into text suitable for an SOP. For further information, refer to one of the many books and Internet articles available on the topic of business process mapping.

Assemble a team

In order to get a complete and accurate process map, it is essential to get all the right people involved in describing that process; creating a team or working group is therefore a critical first step. The lead functional group for the SOP (the business process owner, see Chapter 6) will usually be responsible for initiating the process of forming a working group and may request certain staff, but the management of each functional area will have the final say on who participates. The working group should include representatives for all of the roles that are expected to have an action or task, and the representatives should have some experience with the activities in question. Sometimes, if a role will *only* have review or approval of a document, the person representing that role may not have to sit through the process mapping discussions but can be brought in as needed during review of the SOP that will result. That person can also focus on providing input to the document that will be reviewed.

Whenever a new computer application or system is involved, include someone with technical knowledge of the system in the process team. This person may have to be a representative from the company's IT or system support group, or even a consultant from the vendor or other external source. This person may not have any assigned actions in the planned process but should be able to identify technical issues that place a constraint on a particular step in the process or on who performs a step.

Review regulatory requirements

Any major update to an SOP should start with a review of regulatory requirements and expectations. The requirements set forth in regulations do not change that frequently, but there is no harm in refreshing one's memory or exploring specific requirements as seen in Example 1. What does change frequently are the regulatory expectations as presented in guidance documents from the FDA or reflection papers from the European Medicines Agency (EMA). Draft and final guidance documents provide a strong hint as to matters that concern the health authorities and frequently result in changes in the way clinical trials are conducted. Two very good examples are the FDA's *Oversight of Clinical Investigations—A Risk-Based Approach to Monitoring* and the EMA's *Reflection paper on risk based quality management in clinical trials*, both of which have changed monitoring processes at many companies. No regulations changed but the direction of the industry has.

Example 1

A company using electronic data capture (EDC) was having trouble getting principal investigators to apply their electronic signatures for every subject in the trial before the study database was locked. A large part of the problem was the way in which the electronic signature activity was configured in their implementation of the EDC system. They convened a working group to see what other approaches could be taken or what other configurations were possible. Early on, the idea surfaced to *not* obtain signatures because there was no regulation requiring it! When the process group for one of the functions heard of this idea, they had to prepare a short presentation saying, yes, investigator signatures on CRFs are required as per *ICH E6 GCP*[*] Section 8.3.14, where "signed, dated, and completed case report forms" are listed as essential documents. This occurred a few years ago, before the recent FDA guidance documents on electronic source documents, which emphasizes the importance of having investigators sign to approve and release the data in electronic case report forms (eCRFs) and provides additional information about the expectations of signatures on eCRFs.

List steps or actions in the right order

Once the working group has been established and the goals or problems to be solved have been set, the group starts by listing five to ten of

---

[*] International Conference on Harmonisation, *Guideline for Good Clinical Practice*, E6.

the high-level steps for the process. The group members should avoid going into too much detail until the high-level steps have been agreed to. Then the group goes to the next level of detail for these actions. Chapter 4 discussed the level of detail appropriate for SOPs and recommended a medium level of detail that lists the high-level steps to identify *what* and then includes detail (or the *how*) when there are required tools, hand-offs, or approaches that are needed to insure quality or efficiency.

If the high-level process comes to more than nine or ten steps, consider splitting the main process, which may or may not result in a second SOP. Multiple maps can still be accommodated in SOPs as they typically have different sections for different processes, but the group must be mindful of keeping the SOP scope and purpose to a concise unit (see Chapter 4). Similarly, if a high-level step is complex enough that when broken down it warrants its own map, this situation can still be accommodated by a single SOP unless it is true of a majority of the steps. Note however that, when there are different ways of performing a step but they all result in the same outcome and quality, it should not be broken down further. This will allow for flexibility to be built into the SOP. Acceptable options for steps in an SOP and the details of how an action is to be performed are best left to supporting documents or training.

Here are two effective methods for identifying the steps or actions and their order:

- Use sticky notes on a whiteboard or wall. Write the key steps on the notes and move them around until they are in the right order. This technique is particularly useful for processes that are not strictly linear, because the notes can be placed in any arrangement. Photograph the result as documentation and as the source for the action items.
- Project a Microsoft Word document in the outline view. The outline view allows the main steps and any associated sub-tasks to be moved around very easily. This method is conveniently self-documenting and provides the shell of a process document. This option is especially useful when the working group is not able to meet in person.

There are other software applications available to facilitate brainstorming that may also be helpful. Avoid using a flowchart program (such as Microsoft® Visio) during the initial mapping sessions because it usually results in a lot of time being wasted as the person on the keyboard works through the software. After the steps have been drafted using informal methods, a member of the group familiar with the flowchart software used by the company can translate it into the appropriate process map for review by the working group.

Who is responsible? Who is involved?

During the discussions around the high-level steps of the process, the working group likely will have discussed who should perform a particular action or who oversees a task. If that has not happened already, it should happen next. For each step in the flowchart, the working group must identify who is the person *primarily* responsible for the outcome of the step. This process can be challenging when the step is a high-level action that summarizes more detailed steps and can lead to splitting steps in the process to make responsibility clear—as it will need to be in the written SOP. In best practice, each step has a single responsible role associated with it, though there will be cases where multiple roles are performing the same step (perhaps in different ways) to ensure a high level of quality. As we see in Chapter 9, some companies consider review or approval of an action or document as an example with multiple responsible parties; others require that a single role take the lead in ensuring the review and approval occurs. Compare this with user acceptance testing of an electronic case report form—although several roles may play a part in testing, they are not all equally responsible for the outcome. Usually a single role, often a data manager, is responsible for overseeing all of the testing and for documenting and summarizing the results.

The working group should also identify the roles that perform some of the subactions of a step, provide consultation to decisions made, or provide content for inclusion in a document. Whereas there may be heated discussions around the ultimate responsibility for an activity, input or involvement in an activity generally raises fewer issues but it should still be taken seriously. Chapter 4 pointed out that the *who* of an SOP has an impact on the resources of a functional group; this is the time to point that out to the working group representatives. Because the SOP that results from this mapping process will act as a kind of contract between functional groups, the working group team may have to take the questions of responsibility and involvement back to their functional management before the process is finalized. The responsible and involved roles will also impact training, as we will see in Chapter 15.

In some flowcharting programs, the key responsible role can be added to a workflow diagram at the bottom of each process box. These diagrams focus on the process—what steps or actions are being performed. For workflows that focus on responsibility, *swim lane diagrams* (also known as Rummler–Brache diagrams) are used. In swim lane diagrams, responsible roles are assigned to horizontal or vertical columns, and the process steps are arranged in the lane or lanes that apply. These diagrams can identify areas of high risk should communication or hand-offs fail, but because the arrows can be all over and because items with multiple responsibility can be hard to indicate, the diagrams can be quite complex. Figures 8.1 and 8.2 both show the same process flow for identifying the visits that will be used in the EDC system for site data entry, central lab vendors for data transfers, and the clinical trial management system for tracking visits and payment. In this example, based on an actual process, the study team lead, data manager, lab data manager, and EDC programmer review the visits in a specific order, then only the study team lead, data manager, and EDC programmer approve.

When using either the swim lane or the process flow approach, it is helpful to be aware of the meaning of standard shapes and use standard approaches, but in the end the goal is to create a good visual representation of the process that adds to understanding. That may mean making changes to what is standard for a process diagram. As the diagrams are created, the process team may ask themselves, "What is important about this process; does the diagram show it?" Be prepared to try unusual approaches, as in Example 2.

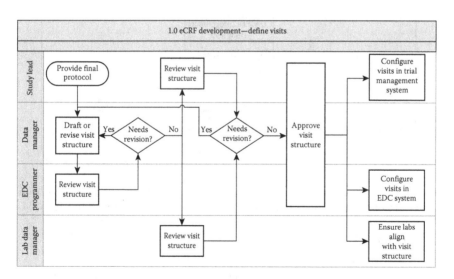

Figure 8.1 A swim lane diagram for defining the visit structure early in the electronic case report form design process. The focus is on the roles involved in the process and when they are active.

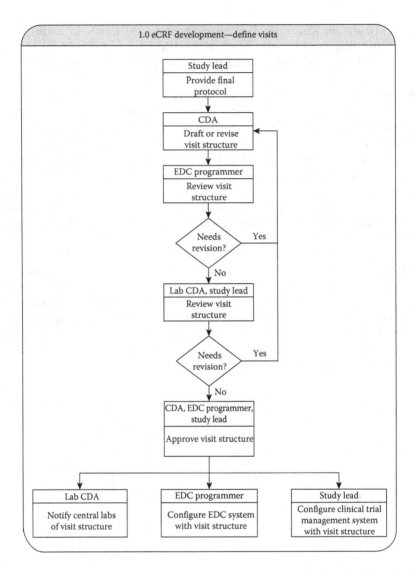

Figure 8.2 The same process using a simple flowchart where the roles have been added. The process appears to be less complicated but is the same as in Figure 8.1.

Example 2

A smaller company wanted to map clinical development activities for the first time, to provide some consistency across their growing number of studies and investigational products. They hired a consulting firm with professionals experienced in both facilitating mapping and presenting the results. The consultants used proven methodologies at first—starting with workshops where participants used sticky notes on large pieces of paper. But after the workshop, they were able to present the information in new ways using mapping software. One process was particularly challenging: the company wanted to recognize the importance of creating the many plans associated with a clinical trial—including those for safety management, subject recruitment, site monitoring, and data management—by putting it in its own diagram. In the process mapping workshop, the participants identified the plans, the input documents or information they required, and the start and end milestones, but it was not a coherent process flow. After the workshop, the consultants made sense of the process for creation of the plans by creating a diagram that had just one process box for each plan, and then grouped the plans by common inputs and highlighted the milestones—not a typical approach but it made the diagram and its purpose easier to read and to evaluate in the context of the other maps.

Identify inputs and outputs

Example 2 showed how inputs to a map can be used to both organize elements of a process diagram and check it against other maps in the overall clinical development process. In some mapping philosophies or approaches, each step in a process diagram has to have an identifiable input and an output (that you can count) and these items are noted explicitly on the process diagram. Although adding a step-by-step input and output may only help a bit in clarifying a process, listing the *overall* inputs and outputs from the process can be very helpful in identifying gaps at the broader Clinical Development level. In the maps in Figures 8.1 and 8.2, the final protocol is an input, which means that this map for defining the visit structure could not take place until after the protocol is finalized. The key output from the maps is the approved visit structure. That visit structure is used to configure three different systems for the study in question: the database used by the central lab, the EDC system used to collect site data, and the clinical trial management system. The configuration of the clinical trial management system for a study may well be a complete process of its own (perhaps with an SOP) and it is important to understand for that process where the information comes from and when in the study startup period the information will be available.

Be aware of sequence and prerequisites

The mapping of the process will provide the *who* and *what* needed to write the SOP and enough of the *how* to provide sufficient detail to ensure an appropriate level of quality (see Chapter 4). Although we touched on the *when* in discussing inputs and outputs, there is more to consider.

If it is not otherwise specified, a linear process map implies that each step is completed before the next one can begin. When a map is translated into an SOP, the same implication carries forward to the text. There are plenty of examples of good clinical practice processes where work can be carried out in parallel, but then must be completed in a particular order. In Clinical Operations, there are several threads of activity that are related to site initiation but many have to be complete before subject enrollment begins. Also during study startup, a study team may start drafting an eCRF, the associated edit checks, and the CRF completion guidelines, all in parallel. However, the CRF must be finalized and approved after the final protocol and before the edit checks and the CRF completion guidelines can be finalized and approved. Parallel paths can be incorporated in process diagrams and prerequisites can be added to process maps by using appropriate symbols. Add more complex conditions such as "before enrollment of the first subject" or "no later than three weeks after study database go-live" as comments or footnotes to a process step so they are not lost when the SOP writer translates that step.

Translate the process into an SOP

Once the working group agrees on the process, the members may decide to have other people from their functional groups review the process diagram before drafting the SOP. When the process looks correct, the designated SOP author (from the working group or a new person) will take it over and translate the steps using the company's SOP template. Process diagrams translate more easily to templates that use active language and tables showing clear responsibility (see Chapter 9), but everyone involved will be surprised to discover how many issues are brought up by the translation from diagram to text. The working group should be prepared to continue to be available to resolve the new issues that will arise during drafting and in the SOP review that follows.

Current practice at many companies is to require an SOP to include at least one process diagram. Other companies encourage the diagrams as appendices. Even though the SOP text was likely written from the diagram, there can be some minor difficulties in aligning the two, because the diagram and text are trying to convey different messages. They may appear not to match, such as when the SOP text includes multiple steps that are all part of one "box" in the process diagram. Or, the diagram may be better able to convey the order or cycling of steps than the text. It can be helpful to set expectations by labeling the diagram or adding text to indicate it is one "view" of the process. Even when the diagram is physically in the same file as the SOP language, the text and diagram can diverge over multiple revisions because SOP authors commonly forget to review the diagram after revising the text following a review cycle. A checklist that includes an item to remind the department process group or the Controlled Document group to review the diagram works well to address this problem. The scenarios described in Example 3 show some of the other things that can go wrong with including diagrams—but these are not meant to imply that diagrams are a bad idea, merely that there are issues that may arise and it always helps to think ahead.

Example 3

Because creation of a process map prior to drafting an SOP is considered an industry best practice and because many staff members benefit from the more visual representation of the process, some companies have made inclusion of a process map a requirement for SOPs. One company that tried requiring the process diagram as the first section of each SOP, prior to the text of the procedure, ran into unexpected problems and rescinded the requirement after a short run. The process mapping software that they had intended everyone to use did not work as promised, so different functional groups ended up using different process mapping software. Also, different authors of the diagrams had different styles of mapping. Some authors interpreted the requirement of a process diagram as meaning every step in the SOP had to be reflected in the process diagram, making the diagram overly complex. Other groups summarized or grouped steps to provide a high-level summary, but this caused confusion about whether the diagram (which implied flexibility) or the text (which was prescriptive) was the rule to comply with. Another company that had run into this very problem addressed it by requiring that all processes be mapped by trained specialists in the Controlled Document group—although this introduced quality and consistency to the process diagrams in SOPs, it added several weeks to the timelines for release of an SOP!

Stay aligned with any approved process maps

At some companies, the diagrams that come out of the working group will be a formally managed document and may even require approval from function managers (see Chapter 2). In this case, it will be essential to ensure that the diagram for the SOP and the officially approved version of the process flow are always aligned with each other. If writing the SOP text highlights a discrepancy in the approved diagram, the formal diagram should be updated. At one company, the group managing the approved diagrams was keeping a list of changes to make "in the future" as the SOPs for the constituent maps were drafted and identified needed changes to the approved maps. This divergence of the approved diagrams and diagrams in the SOPs caused significant confusion among staff who were referring to both the SOP and the process diagrams to try and understand the new processes being implemented. This cannot be allowed to happen.

SOP of SOPs

Most SOPs of SOPs will not explicitly require a process diagramming step for creation of an SOP, but the requirement can still be enforced through the SOP template. A manual for SOP writers can provide guidance on styles (swim lane or process flow), software, and choices such as dealing with multiple responsible roles. When a company requires mapping by the Controlled Document group (see Example 3), this could be reflected in the SOP of SOPs or it could be considered part of a more general formatting step—but the time to allocate must be made clear to authors.

chapter nine

The SOP template

Having mapped the process to be documented by following the recommendations in Chapter 8, we are ready to translate it into text using an SOP template. The SOP template is more important than might be initially apparent because it embodies key aspects of a company's SOP philosophy and provides more detailed guidance than is possible in the SOP of SOPs. The structure of the SOP template and the format for each of the sections also can have surprising implications for compliance. This chapter lists the key sections of an SOP and provides guidance for the content of those sections. When reading this chapter, refer to the example SOP template found in Appendix 1, which was designed to demonstrate the recommendations that follow.

Document header

The content of the document header for an SOP can provide vital information to the reader. The SOP's title, identifying number, version, and the page number would seem to be natural choices for information to appear on each page of the document. The example in Appendix 1 does include all of these for the reasons described below. However, some companies choose not to include a full header after the first page. They make this choice in an effort to include more text of the procedure on each page, but in doing so they remove an important anchor for readers scrolling or flipping through the procedure. For example, one company did not include the title of the SOP on each page, only the document number. Because most users do not remember which ID numbers go with which SOPs, they have to refer back to the first page to get the full title, and as we will see below the title provides important scope information.

Title

The title of an SOP should be both relatively short and yet quite specific. There is no need to use "An SOP for..." or similar language as part of the title. The SOP title, like the procedure itself, should use short, concise language, for example:

- "Developing a Protocol" or "Protocol Development" not "An SOP for Protocol Development"

- "Creating and Maintaining the Statistical Analysis Plan" or "Statistical Analysis Plans" not "How to Create a Statistical Analysis Plan"
- "Processing Serious Adverse Events from Clinical Trials" not "The Process for the Entry and Reporting of Serious Adverse Events Identified in Clinical Trials"

The title is essentially a summary of the *Purpose* of the document.

When a title includes a phrase that is associated with a very common initialism (or abbreviation), include it, because when users perform searches within SOP titles they are most likely to start with the shortened version of a phrase and will only spell it out if nothing is returned in the search. So for an SOP on testing a study developed using an electronic data capture system, consider a title like, "Testing Electronic Data Capture (EDC) Studies," which has both the long and short form of EDC in the title. Companies should standardize this approach of using both forms because the biggest danger arises when there is a mix. One company had these SOP titles: "Developing Electronic Data Capture Studies" and "Testing Electronic Data Capture Studies," but they also had: "Study Closeout for EDC Studies." Searches in title text turned up one set or the other but not both.

The example template in Appendix 1 has the title appearing only in the header rather than having it appear in title font on the first page. Including the title on the first page does not add value if the header is the same across all pages; in that case, it really only takes up space. With a different header for the first page, it becomes a stylistic preference.

Document identifier

The identifier or document number is a very important concept in the control of documents. Documents must be uniquely identifiable. Although in the good clinical practice environment a title typically can identify the document uniquely, most companies also assign document identifiers. The advantage of making the identifier the unique key to the document is that it allows the title to change during future revisions. This is a fairly common event over the life of a document and often reflects changes in the scope of the contents. A detailed discussion of controlled document numbering and a proposed numbering scheme are found in Chapter 14.

Effective date

One item often overlooked in designing a header is the approved or effective date of the document. That date provides an important piece of information for the reader: is this a recent SOP or an older SOP? Although we may wish otherwise, it is the case that the longer an SOP has been effective without revision, the more likely it is to have sections that are out of date. When new employees review older SOPs as part of their training, they are often confused by slight differences in role names or changes in tasks that, although they do not make the SOP completely invalid, do not completely match the current state (see also Chapter 12). Knowing the approval or effective date puts the procedure in context—was it written when the system was just installed? Was it approved before the merger? Does it reflect the functional groups names from the last reorganization?

Some companies have a process for releasing SOPs in such a way that the exact effective date is not known when the SOP is finalized, and so they do not include it in the header. There is so much value to easily knowing the effective date while reading the document that, for these companies, it would be worth setting a fixed effective date a week or two further out to ensure that all the prerelease steps can be completed before the predetermined effective date.

Version number

In the same vein as the effective date, the version number of the SOP provides useful information—though less than the effective date provides. Some companies append the version number to the identifier, so that an SOP with a document identifier of CO-001-SOP and a version number of two becomes CO-001-SOP.2. Other companies keep the version number separate, and this is the option used in the example template in Appendix 1. (Controlled document numbering is discussed in detail in Chapter 14.)

Purpose and scope

The *Purpose* of an SOP is a short description of the content of the document, elaborating on the title. The *Scope* describes the conditions under which the procedure applies. Most companies separate the two into their own sections, but it may not always be clear to an SOP author whether a particular piece of information about applying the SOP is *Purpose* information or *Scope* information. It can help authors if they are given a firm distinction for procedures related to running clinical trials: *Purpose* summarizes the activities in the procedure and *Scope* identifies the types of trials for which the procedure must be followed. The distinction between the two probably does not matter too much as long as the information is clearly stated and available at the start of the document. The example SOP template in Appendix 1 separates *Purpose* and *Scope*, but a strong argument could be made to fuse them into a single section.

Purpose

Information about the purpose of the procedure can usually be stated in a few sentences. For an SOP on developing a protocol, the *Purpose* might read, "This standard operating procedure describes the process for developing a protocol from draft through approval. It specifies the reviewers for each stage and identifies the necessary approvers." When describing what activities are not covered, the *Purpose* might refer to other SOPs as an aid to readers: "Refer to CO-011-SOP, Protocol Amendments, for the procedures on making changes to a protocol once it has been approved." An SOP on unblinding of single subjects for reporting serious adverse events to regulatory authorities might read, "This standard operating procedure describes subject unblinding for purposes of adverse event reporting. It includes the process for authorizing the unblinding of treatment for a single subject and explains the steps to ensure that the study team remains blinded. Refer to BM-104-SOP, 'Single Subject Unblinding for Emergency Treatment,' for the procedure to be followed in providing treatment assignment information to investigator sites."

Scope

In terms of its impact on compliance, *Purpose* is not nearly as important as *Scope*. When well stated, the *Scope* of the procedure can aid correct implementation, and poorly worded *Scopes* can result in unnecessary deviations or improper use. If we focus, as proposed earlier, on the kinds of trials to which the procedure applies, then we can begin to narrow down the *Scope* through a series of questions. Examples of those questions to determine *Scope* include whether the procedure in question applies to the following:

- All trials for which the company is the sponsor or only those trials for which the sponsor conducts the trials (i.e., they are not outsourced)
- Interventional trials or also noninterventional trials
- Blinded trials only or also to randomized trials that are not blinded
- Paper-based trials, EDC trials, or both (or any other system dependency)

A brief discussion of each of these will show some of the subtle implications to consider. The first question tries to clarify whether a procedure applies when the trial is conducted by the company's staff and contractors or if it also applies when the activity is performed by a contract research organization (CRO). For example, when data management is fully contracted out to a CRO to be conducted on the CRO's own computer systems,

CRO staff would typically follow the CRO's own SOPs, not the sponsor's. So a sponsor's study build SOP would apply to in-house studies only. However, even when data management is outsourced to a CRO, there will be at least a partial study team at the sponsor, and at most companies this team will perform some kind of data review on the CRO's data. Therefore, the SOP on data review, which typically requires the creation of a data review plan of some kind, would apply *both* to studies where data management is conducted in-house and to studies where data management is conducted by a CRO. Finally, any SOPs created to guide oversight of CRO activities obviously apply to activities fully outsourced to a CRO, but does the oversight SOP also apply if a CRO provides staff to work on the sponsor's systems according to sponsor SOPs? The latter is sometimes called a *functional service provider model* and the SOP may or may not apply.

The second bullet focuses on whether SOPs apply to interventional or noninterventional studies (NIS).* At some companies, NI studies are run by a division of the company separate from that conducting Phase I–IV trials; at others, some or all of the trial-related activities may be performed by the same group that conducts interventional studies. In companies where some or all of the NIS activities are conducted by groups in Clinical Development, we quickly run into issues of different structures of study teams, different requirements for data cleaning, and different expectations for monitoring, all of which must be properly reflected in SOPs that apply to both kinds of trials. At one company where noninterventional trials were run by the same division of the company as other trials, they attempted to word expectations and roles to allow flexibility wherever they could, but it was a challenge that many of the SOP reviewers did not understand, and the company created a group to review all SOPs in development or revision to assess the language and requirements against the needs of noninterventional trials.

Because SOP authors often come from regular staff, they may not be aware of the attributes of trials outside of the ones they have been working on—if the author has been working on later stage trials, that person may not be aware of some of the particular features of the early stage trials. The third bullet attempts to get clarity about one such difference. If an author works primarily on later stage trials, in that author's mind "randomized" may be equivalent with "blinded" and randomized, open-label trials may not be dealt with correctly in an SOP. See Example 1 for a *Scope* that led to SOP deviations being filed because of this distinction.

---

* Noninterventional studies are those where the treatment is prescribed in the usual manner in accordance with the marketing approval. Noninterventional studies include postmarketing surveillance studies, postauthorization safety studies, and other kinds of observational studies where patients are treated in real-life conditions.

Example 1

A company had one SOP for subject numbering in clinical trials that used Integrated Voice/Web Response Systems (IxRS) and a different SOP for studies that did not use IxRS and so required manual assignment of subject number blocks for each site. When the SOP on manually assigned subject numbers was written, the *Scope* was written to cover subject numbering for clinical trials that were "not randomized" because the SOP author assumed that nonrandomized trials would not use IxRS. Soon thereafter, IxRS began to be used at this company for many nonrandomized trials as a means of monitoring drug shipments to the sites. When a review of the trial master file for a group of nonrandomized studies identified documentation of manual subject number blocks as missing, the company's compliance group was asked if a deviation was required. The compliance group ruled that until the SOP could be revised an SOP deviation would have to be filed for each study that was not randomized but still used IxRS, because the *Scope* of SOP for manual assignments clearly said "not randomized." Also, retrospective deviations were required for studies that had inadvertently ignored the requirement in the past. (It is worth noting that compliance groups at other companies might not have decided it this way, because the *intent* of the SOP for assigning subject number blocks was that it be used for *manually* assigned subject numbers; whether or not the study was randomized was actually not the issue.)

The *Scope* section provides the right location to specifically call out whether a procedure applies only when a certain software application is used, as is shown by the final bullet. A procedure can be impacted by a software application when the application imposes a certain order to steps or introduces steps that might not occur if another system had been used. System dependencies still appear most frequently in clinical data management SOPs (see Example 2), but as all groups in Clinical Development move to software specifically written for their functions, it is showing up in other areas including clinical trial management systems, electronic trial master file applications, and specialized systems to support data analysis. If the *Scope* of an SOP limits its application to studies that use a particular system, it may be worth reflecting that in the title as well as in the *Scope*.

Example 2

A company had a general SOP—which could almost be called generic—to cover key aspects of data management for EDC systems during a time when they used multiple, vendor-supplied systems for their trials. When they decided on a system to be used for all trials going forward, they began to develop SOPs with procedures tailored to that system. At first, the generic SOP and specific SOPs were in sync, so that the general requirements matched the specific requirements, but as time went on, the company wanted to move in specific directions with the new EDC system and optimize processes for it. The original, generic EDC SOP said it applied to trials using electronic data capture, so to allow different procedures in the system-specific SOP, the *Scope* of the original SOP had to be revised to apply only to the EDC studies conducted in legacy, or predecision, EDC systems.

If the *Scope* does not specifically say otherwise, an SOP would be assumed to apply to all trials in which the activities in question take place—with the exception of outsourced activities, which would be assumed to follow the vendor's SOPs unless otherwise stated in the scope of work for that project.

Definitions and background

Every company should create and maintain a corporate glossary for words, phrases, acronyms, and initialisms that commonly appear in controlled documents. Chapter 19 proposes approaches to creating and maintaining glossaries and suggests that if a term needs to be defined in two documents it should be added to the glossary. Terms, acronyms, and initialisms that are common to the industry can be added to the glossary from the start. Current practice is to not list terms in the *Definitions* section if they can be found in the glossary; the reader of the SOP is expected to try the glossary when encountering an unfamiliar word or phrase. Having a hyperlink from the *Definitions* section to the glossary is very handy and encourages readers to look up terms.

Companies differ as to whether acronyms and initialisms common to the industry, like CRF, EDC, PRO, and CSR, which may be deemed familiar terms or are found in the glossary, should still be spelled out at the time of their first use in every document. Some require it and others do not. Certainly other less common terms that do not need to be defined but for which it is convenient to have a shortened form should always be spelled out at first use, for example: "principal investigator (PI)." These do not appear in the definitions.

The focus of the terms that do appear in *Definitions* should be to explain slightly unusual usage or usage very specific to the controlled document. Definitions are especially important when a company's SOP template does not support a *Background* section as described below. In that case, the *Definitions* tend to be a bit longer and more terms are included to ensure that tasks in the *Procedure* sections are clear to all readers. For example, at a company where the SOP template did not permit background text, an SOP on processing completed subject case report form (CRF) PDF files for inclusion in a submission included the precise meanings of "Blank CRF," "Completed Subject CRF," and "Processed Subject CRF" in the *definitions* to ensure that all readers understood both the important differences in the terms and how all three fit into the overall process.

The *Background* section, if available, should be used for providing any context that aids the reader in understanding the whole of the procedure steps that follow. Remembering that the audiences that pay the most attention to SOPs are new employees and auditors/inspectors, *background* can provide specifics to the process related to the procedure that might not be immediately understood or obvious to someone unfamiliar with the company. There is no need to repeat information that will become clear in the procedure that follows, but when the *Background* section is used judiciously, it adds clarity to the procedure as it allows for more natural wording than the format of the *Definitions* section does. If in the example above for CRFs for submission, the SOP template had permitted background or context information, then the author would have provided a paragraph explaining each of the types of CRF PDF files and when they are produced. The information would be roughly the same whether it appeared in *Definitions* or in *Background*, but there are cases when a definition cannot provide the necessary context, as seen in Example 3:

Example 3

A company was using an EDC system that had a complex set of steps to implement database changes after the study had gone live. The database structure was *amended* and then the data was moved or *migrated* to the new structure. A *mapping plan* showed how the original data was to be moved to the new structure and how edit checks were to be executed during the move. For studies with a large number of sites, data could be migrated in a *gated* way to allow sites that had approved the protocol changes to which the database amendment was associated to receive the new electronic CRFs (eCRFs), while those sites that had not approved the changes continued on the old structure.

To explain this context using definitions, one would have had to define the terms *amendment, migration, mapping plan*, and *gated migration*, and these terms would have appeared in alphabetical order in the *Definitions* section, which would have made the explanations confusing. Because the company's SOP template in this instance permitted a *Background* section, the SOP included a paragraph, similar to the first paragraph in this example, to explain the special terminology concisely and clearly.

Needless to say, if a term appears in the *definitions*, then that term must actually be used in the SOP body. Authors will sometimes add definitions for terms that appear in early drafts or versions of the procedure and then forget to remove the definitions when the text is later changed and the terms are no longer used.

Responsibility

All too often, a separate *Responsibility* section in an SOP simply leads to authors repeating parts of the procedure found elsewhere. That repetition or duplication of information has little value and introduces room for error when the summary found in the *Responsibility* section and the actual procedure do not exactly match, as often happens during repeated edits of a draft document. Even when SOP authors succeed in concisely summarizing the responsibilities found in the body of the procedure, it does not add value to the document as a whole—because the information is there in the next section already. A *Responsibility* section that simply rephrases responsibilities that follow in the procedure should be dropped; however, when it helps readers quickly identify the roles that have key actions in the procedure, then the *Responsibility* section can add value to an SOP. It is the focus on the impacted roles, rather than a restatement of procedures, that is the difference.

One useful format for highlighting key roles in the *Responsibility* section is a *RACI* (*responsible, accountable, consulted, informed*) table. The *Accountable* role, and there can only be one, is ultimately answerable for the outcome of the entire procedure. The *Responsible* roles actively perform the tasks or actions to complete the procedure. The *consulted* roles provide information or input on request. *Informed* roles only receive information; the communication is one-way. (Because of the importance of the accountable role, the table is sometimes called an ARCI.) When a single accountable is identified and key responsible parties are listed, then this information can be used to determine necessary training. Accountable and responsible roles always have to train; those roles that are consulted or informed do not have to train on the procedure, but they may want to (see also Chapter 15).

A simpler form of the RACI approach, which still adds value to the document during review and for training, is to simply list roles in a *Responsibility* section that have active tasks in the procedure and those roles that only review and approve actions, information, or documents. Companies that do not require a single accountable party for the full procedure may find this latter approach both useful and flexible. This simplified approach to the *Responsibility* section is the one used in the example template in Appendix 1.

In Chapter 4, we saw that the roles listed in *Responsibility* may or may not be the same as a job title. For example, *data manager* may be both a title and role, but *study leader* may exist only as a role for purposes of writing SOPs. These special roles can be defined in the *Definitions* section or the corporate glossary. In either case, they must be associated with a specific department or function in the organization so that training can be appropriately assigned. For example, if *study leader* is not a job title but a role, then the definition must specify that staff performing in this role come from the Clinical Operations department with a job title of, for example, either clinical project leader or clinical trial manager.

Never name an entire department or function as a responsible party either in the *Responsibility* section or in the *Procedure* unless it is actually true. For example:

- The study data manager is accountable for study database build specifications, not the entire Clinical Data Management (CDM) group and definitely not the Biometrics department to which CDM belongs.
- The clinical coder is responsible for coding adverse events from clinical trials, not the entire Drug Safety group to which coders belong.
- The site monitor is responsible for obtaining principal investigator signatures, not Clinical Operations.

SOP authors may try classifying the responsible role broadly to build flexibility into the process, but it really just introduces vagueness in the true responsibility and can add a large training burden to all functions. If the clinical coder is responsible for coding adverse events, then the safety associates who enter adverse events into the safety reporting software application (and who also report into the Drug Safety department) do not need to train on the SOPs that guide coding activities unless their management determines that it would be to their benefit.

Procedure

In the *Procedure* section, the SOP author takes the process maps developed (or modified) as a first step in defining a new process, as described in Chapter 8, and translates them into text. The *Procedure* section may have multiple subsections, with one for each major activity in the process map.

Outline format

One common format for the *Procedure* section is an outline style. Using the process diagrams from Chapter 8 found in Figures 8.1 and 8.2, the text for that portion of eCRF development could be written as shown in Example 4.

Example 4: Process from Figure 8.1 translated using an outline format

6. Procedure

6.1 Define Visits

The visit structure for the trial is finalized as the first step in eCRF development because the visit names permit integration between the Clinical Trial Management System (CTMS), the study database, and the statistical analysis software (SAS) programs used to manipulate data provided by central labs.

> 6.1.1 The Data Manager receives the final study protocol from the study lead and drafts the visit structure using the visit structure (VS) template and sends it to the EDC Programmer for review.

> 6.1.2 The EDC Programmer reviews the VS document and works with the data manager, who revises the VS document as needed until both agree that the VS document is ready for further review.

> 6.1.3 The Data Manager sends the document for review by the study lead and the lab data manager and incorporates their feedback.

> 6.1.4 The Data Manager repeats step 6.1.2 and 6.1.3 until no further revisions are needed. The Data Manager approves the document and sends it to the study lead and EDC Programmer for approval.

> 6.1.5 In parallel: the EDC Programmer uses the VS document to configure the study database visits, the study lead uses the VS document to configure the CTMS with the study visits, and the Lab Data Manager uses the VS document to develop data transfer specifications with the central lab.

Table format

The use of the outline format tends to lead to wordier paragraphs where it may be hard to identify who is doing what without a careful reading. Many companies have switched to a more active table format where the responsible parties are called out and associated with the description of the task. Compare Example 5, which uses the table format, against Example 4 in the outline format. The template in Appendix 1 uses the table model.

Example 5: Process from Figure 8.1 translated into a table format. Notice how the procedure is longer but is easier to read

6. Procedure

6.1 Define Visits

The Visit Structure for the trial is finalized as the first step in eCRF development because the visit names permit integration between Clinical Trial Management System (CTMS), the study database, and the SAS programs used to manipulate data provided by central labs.

Step	Responsible	Action
6.1.1	Study Lead	Provide final protocol to DM.
6.1.2	Data Manager (DM)	Draft or revise the visit structure using the visit structure (VS) template. Provide the VS document to the EDC Programmer.
6.1.3	EDC Programmer	Review the VS document and provide feedback to the DM; if no further revisions are required, continue with step 6.1.4.
6.1.4	DM	Provide the VS document to the study lead and lab DM for review.
6.1.5	Study Lead Lab DM	Review the VS document and provide feedback to the DM.
6.1.6	DM	Incorporate feedback; repeat steps 6.1.3 through 6.1.5 until no further changes are required. Approve the VS document and provide it to the study lead and EDC Programmer for approval.
6.1.7	Study Lead, EDC Programmer	Approve the VS document.
Steps 6.1.8 through 6.1.10 are performed in parallel:		
6.1.8	EDC Programmer	Configure the study database visits.
6.1.9	Study Lead	Configure the CTMS with the study visits.
6.1.10	Lab DM	Develop data transfer specifications with the central lab using the study visits.

When using the table format, some companies consider approvers of documents as all being equally responsible. Step 6.1.7 in Example 5 shows the study lead and EDC programmer explicitly as responsible and active participants. At other companies, approving a document is viewed as akin to consulting, and the procedure would be written as shown in Example 6, with no separate approval step and moving directly to configuration.

Example 6: When the approval step is implied

No.	Responsible	Action
6.1.6	DM	Incorporate feedback; repeat steps 6.1.3 through 6.1.5 until no further changes are required. Approve the VS document and provide it to the study lead and EDC Programmer for approval.
6.1.7	EDC Programmer	Configure the study database visits.

Example 7 shows the implications of this approach more clearly. Compare the first table in Example 7, where the responsibility of the medical monitor and clinical coding associate is explicitly called out, against the second table with the more passive approach.

Example 7: Two ways of defining an approval: first actively then more passively

No.	Responsible	Action
6.1.1	Clinical Data Manager (DM)	Send the listing of coded adverse event and medication terms to the medical monitor and clinical coding associate.
6.1.2	Medical Monitor and clinical coding	Review and approve the coding according to CC-001-SOP, "Coding of Terms in Clinical Trials."

No.	Responsible	Action
6.1.1	Clinical DM	Send the listing of coded adverse event and medication terms to the medical monitor and clinical coding associate for review and approval according to CC-001-SOP, "Coding of Terms in Clinical Trials."

Both of the approaches in Example 7 are acceptable, but the critical difference is that in the first case, the roles in 6.1.2 are considered responsible for an action and so they would normally be asked to train on the SOP. The thought here is that these roles need to be aware that they have to perform this activity. Also, as we will see in Chapter 10, the Clinical Research group and Drug Safety group (to which Coding belongs, perhaps) would also be asked to review and approve this particular SOP because their groups have responsible actions. When the approvers of study documents are listed passively or as consulted, then they have more flexibility in the training. They may request training but are not required to train, and they would not be asked to approve the SOP. When adopting this consulting approach, it would be appropriate to insist that consulted roles be included in the SOP's cross-functional review so that they can agree or disagree with the assessment that their role is to approve a study document. The example SOP template instructions in Appendix 1 uses the approach of splitting out reviewers and approvers into a responsible role and this meshes well with the use of a *Responsibility* section that highlights review and approval as a separate kind of responsibility.

Roles that cannot appear in procedures

One hard and fast rule about the roles in procedures needs to be mentioned here: a role cited as having an action in the outline format or appearing in the Responsible column in the table format cannot be someone from outside the company; in fact, it cannot even be someone within the same company that does not use the same controlled document system. The first is more common, and SOPs governing transfer of clinical data from vendors such as central laboratories frequently run into exactly this issue. Data transfers of clinical data generally require some kind of data transfer specification and a secure transfer process. There is typically an SOP for this process and the vendors providing data electronically are involved in several of the steps, but no step in that SOP can have the vendor as the responsible party, because the vendor does not train on the SOP and is not held to its provisions. In these cases, the step is generally written with the focus on the internal group with steps such as "Wait for notification that electronic data has been received from a vendor" for the group that receives electronic transfers and "Request review by the vendor" for the writer of the data transfer specification. Any actions required of an outside party, such as review and approval of a data transfer specification or use of particular means to transfer data, must appear in the contract with the vendor to ensure they are funded—another example of how SOP steps have an impact on resources and budgets.

Another role that should not be used in procedures is that of *designee*. For a while, it was very common to explicitly say that work could be delegated to someone else in all those cases where it could be by saying "or designee" and to leave it off when it had to be the role in the procedure. For example, one might see "study leader or designee" for tasks which could be delegated to someone else but only "study leader" when it required that specific person to perform the task. The use of *designee* has gone strongly out of favor and most controlled document groups now say never to use the term. Always assume that most work can be appropriately delegated. However, Regulatory Compliance groups have given the guidance that approval tasks have to be performed by someone who has the same or more experience as the role listed. For example, in the case of a document requiring approval by a clinical scientist (medical monitor), a Regulatory Compliance group said the clinical science associate, which was a job title with less experience, could not approve that document even if the Clinical Science associate reviewed the document.

Document disposition

As discussed in Chapter 5, most SOPs currently do not include a specific and clear section on the filing and retention requirements for the documents, forms, reports, or other output that is created as part of the SOP procedures. In Chapter 5, we saw that the study's trial master file (TMF) is one important destination for output from SOPs, but not all documents should be stored in the TMF. Some may be stored in study files for the duration of the trial only; others may be stored for some period after completion of the trial in a central or offsite storage location. Yet others can be disposed of right after the action taken from them is complete, and the action outcome is all the evidence that is necessary.

A *Document Retention* section of the SOP can provide clear guidance to SOP authors by requiring that every document, form, checklist, listing, or other output created by the steps in the SOP have associated with it an ultimate storage location. This section can take the form of a table as shown in Example 8, or it can be a simple bulleted list.

Example 8

Many companies have a study database lock checklist to help ensure that all key data completion and data cleaning steps are completed before study database lock. These checklists have many steps and are designed to have each item checked off and then be signed by the clinical DM, attesting that all steps were performed. Checklists for studies using EDC generally include a step to run a principal investigator signature report to ensure that signatures have been applied to all subjects' data. Most study database lock procedures also have a form signed by key members of the study team when all required data cleaning steps have been completed and no further changes are expected to the data. The document disposition for these three items would look like this:

Document or output	Disposition
Study database lock checklist	TMF
PI signature report (output)	Not retained; can be recreated if needed
Study database lock approval form	TMF

References

SOPs commonly have a *Reference* section and the guidance companies provide on what this section is for varies somewhat. At some companies, the SOP author is instructed to list regulatory references that apply to the activities being carried out as part of the procedure. At other companies, the *Reference* section is the place to list all of the internal documents referenced in the procedure.

The first use does not provide much value to a reader of the SOP, but can help guide the SOP draft and revisions. Chapter 8 recommended that the group creating the process from which the SOP would be written should review regulations and guidance documents as a preliminary step prior to mapping. When this is done, the author can list the most pertinent references in the resulting SOP, if that is company policy. However, the references would have to be more specific than "ICH E6 GCP" or "21 CFR Part 11" in order to have any value at all. At a minimum, a section number from the referenced document would have to be provided to have any real value for either readers or future editors of the SOP. The second approach to the *Reference* section has some value to the reader and a great deal of value in maintaining SOP cross-references; this is the approach used in the example SOP template in Appendix 1 and Example 9 shows how it would be applied.

Example 9: In an SOP on study database closeout, two of the steps might be

No.	Responsible	Procedure
6.1.1	Clinical Data Manager	Sends the listing of coded adverse event and medication terms to the medical monitor and clinical coding associate.
6.1.2	Medical Monitor and clinical coding	Review and approve the coding according to CC-001-SOP, "Coding of Terms in Clinical Trials."

>   Here we infer that CC-001-SOP has steps that lay out the requirements
>   for the medical monitor and coder to ensure adequate review of the
>   coded terms and how approval should be obtained. The author should
>   add CC-001-SOP to the references, and the controlled document group
>   will need to keep track of the connection. Should CC-001-SOP be sub-
>   stantially changed, the SOP on study database closeout would have to
>   be reviewed in light of that change and possibly revised.

One could ask, why is this reference in Example 9 to getting approval
for coding in during study closeout there at all—why not just leave it out
altogether because there is an SOP that covers coding? Referring out to
other documents does indeed require some reflection and is an art rather
than a science. In this case, the coding review is included because it is the
final coding review and must be completed before the study database is
locked. Except for Phase I studies, this will not be the first or only coding
review that takes place in this study; its inclusion in the closeout SOP
emphasizes that a final review is a requirement for lock.

Forms and templates explicitly referred to in the procedure should
also be listed in the *Reference* section. In the example above, if CC-001-SOP
has a form CC-001-FRM-1 that is used to obtain medical monitor approval
for coding, then CC-001-FRM-1 appears in the reference list of CC-001-SOP
but not in the reference list of the SOP on study database lock. The form
attached to the SOP to record final study team approval for lock, say
CDM-012-FRM-1, would be listed in the *Reference* section.

The Study Database Lock Checklist in Example 8 could be a
department-managed document if it includes a lot of details, such as
distribution lists and server names or drop-box folders. Some Controlled
Document groups do not permit department-managed documents to be
listed in the *Reference* section of an SOP, and others do permit them.
Listing a complete set of all types of references is very helpful to both
readers and to those who review cross-references when documents are
updated and so is the recommended approach.

Whenever documents are referenced, users find it very handy to
have a hyperlink to those documents directly in the SOP *Reference* section, and having such a link in a document can improve compliance by
making it easy for users to find exactly what they need to do their work.
Unfortunately, *broken* links can cause compliance problems. If hyperlinks
are going to be permitted, then they must be fully understood and maintained. For example, whether or not links behave properly when a new
version of a document is released depends on the system being used.
If the controlled document system maintains all versions of a document,
when a new version is released does the link go to the previous version or
the new, effective version? If the link leads to the old version, maintaining

those links in all controlled documents that reference it will not be worth the work needed. If the link automatically connects to the new or currently approved version, allowing hyperlinks becomes a viable option. (Note that a few companies have also had problems maintaining links when they changed the controlled document system, but is not a very common occurrence.)

When considering linking references in controlled documents to department-managed documents, the same issues apply, though they are typically worse because the departments are unlikely to be using specialized systems to manage their documents. Department-managed documents are also more subject to changes in company server architecture and the limitations of shared file systems. There have been cases where the entire department-managed document area was moved to another system and all links to those documents broke. As systems for managing controlled documents (and department-managed documents) are improved over time, the link issues may be resolved, in which case, items in the *Reference* section should all be accessible with links.

Appendices

SOP templates typically support a very open and flexible *Appendix* section to permit many different kinds of information to be included. As discussed in Chapter 8, process maps can be included as appendices and diagrams of many kinds can help readers by portraying information in a more visual way. Similarly, tables that summarize information in the *Procedure* section can be helpful in providing another way of organizing requirements. One example of such a use of an appendix is the summary of document types with their review and approval requirements found in the example SOP of SOPs in Appendix 2.

Because appendices of SOPs are flexible, it may be tempting to include more information than is really appropriate. Be aware that screenshots of systems are generally not useful as an appendix of an SOP. Software systems change frequently and screenshots quickly become inaccurate because the effort to revise SOPs for any reason can be high. Inaccurate screenshots can be very confusing to readers, even when they are labeled "example"; this kind of information belongs in a lower level document such as an instruction, manual, or training material that is more easily updated when the software changes. Examples of listings or reports that are referenced in the SOP may fall into the same category as screenshots— they may be tied too closely to systems or include details too granular to be appropriate for an SOP. On the other hand, providing an example can sometimes make the SOP activities much more clear, so the author should weigh the pros and cons carefully.

One rule applies to any appendix: it must be referred to in the body of the SOP. Do not include appendices without references in the body of the document; readers are unlikely to browse through them to see what they can discover.

Revision history

The revision history of SOPs is tracked, but not always in the document itself. Some Regulatory Compliance professionals feel that having the revision history in the document itself can lead to undesirable attention in the case of a regulatory agency inspection. Those companies keep the revision history in the controlled document system, linked to the SOP or work instruction. Many other companies keep the revision history in the document itself, with the thought that knowing what has changed since the last revision and over time is helpful to readers. The example SOP template found in Appendix 1 includes the revision history in the template. Note that the history can have the most recent revision at the top of the version table or at the bottom. Putting the most recent revision summary at the top has the advantage of making it easy to find; but most other types of documents users encounter will have the most recent revision at the bottom, so that may be more familiar. Both are perfectly acceptable.

Other considerations

Some additional notes apply to the management of the SOP template itself, rather than the contents:

Formatting

When the Controlled Document group creates a template, a person very experienced in the text-editing software (generally Microsoft Word) should review the formatting. The headings should use appropriate heading styles and the numbering should be based on those headings to ensure proper increments in numbering when sections are added. If the table format is used, the step numbers also need to increment properly. The indentation should also be preset and be consistent with the heading. SOP authors can waste hours fixing numbering and indentation when the template was not created properly.

Most current SOP templates are in black and white and use the Times New Roman font. This mattered when documents were printed—especially before color printers became common. Now that SOPs are most often viewed on a computer monitor, consider allowing some limited use of color in SOPs. For example, the template in Appendix 1 was created using the default font colors in Word 2010, which has Heading 1 levels in blue. The example also has a prerequisite example row in Section 6.1 where the word "prerequisite" is in red to draw attention to it. These colors are lost in publication but could still add to readability when viewed online.

Template updates

When the template changes, the Controlled Document group has to decide if they want to force all future revisions of existing SOPs to be reformatted to the new template or if they will allow revisions be made that leave the SOP in the previous format. Example 10 shows what can happen when a change to the template is made without strong directions to move existing procedures to it.

Example 10

A company moved from an outline SOP format to a table format. Although new SOPs had to be written using the new template, the Controlled Document group allowed departments to decide whether to reformat SOPs when they were revised or not. Because the table format can make a big difference in wording and intent, reformatting tended to turn up new issues in every SOP where it was undertaken. One common issue was the assigning of responsibility when tasks had previously been referred to in a passive way. A step such as "...was provided to Clinical Operations for distribution to the sites" became "Clinical Operations distributes the document to the sites." Where Clinical Operations previously had not asked to review or approve the SOP, the responsibilities that were now more clearly written warranted review and approval by the group—adding to the time needed to release the revised SOP. This realization led many departments to avoid reformatting longer, complex procedures that would have benefited most from the change. Eventually, even though the Controlled Document group permitted authors to choose the format, the *Regulatory Compliance* group began to insist authors move to the new format.

SOP of SOPs

The SOP of SOPs should require that an approved SOP template be used in authoring SOPs. If the template is stable, it can be a controlled document itself. If it is still in flux, the SOP of SOPs can instruct the SOP author to request the latest template from the Controlled Document group. A manual with instructions for using the template, including some of the guidance provided in this chapter, would greatly benefit SOP writers.

chapter ten

SOP review and approval

If the first draft of an SOP is based on a thoroughly vetted process, the SOP author will not need too much time to translate the process from the process map into a first draft of the SOP text using the SOP template. There may be a few questions of responsibility or wording, but it should be easily accomplished. What will take time—much more time—will be sending out the document for multiple rounds of review and then final approval. Each round of review will generate many comments and issues that need to be adjudicated and resolved before it goes out for the next round. The time needed to solicit feedback and then incorporate those comments is the reason SOPs take a long time to revise.

Concentric rings of review

SOPs need thorough review to ensure that the process can be followed as written, is not missing any key steps, and assigns responsibilities appropriately from the day it becomes effective. Rather than send all the reviewers the first draft, it is best to send it out in waves or rings of review, improving the process and wording as it goes out until the final reviews for approval. Even though the number of people is not necessarily greater in each round, the document goes further out into the Clinical Development organization (Figure 10.1).

The first review is by subject matter experts other than those that drafted the SOP. The second review is by key managers or other reviewers from the functional group sponsoring the SOP, known as the business process owner (BPO). The next level out is a cross-functional review involving all functions or departments that have roles with responsibilities listed in the SOP. The final ring is a compliance group review and a review by the approvers. All of these reviews are discussed in more detail below. At some companies, the functional group acting as BPO manages the entire review process, which involves notifying the reviewers, sending out or linking to the document, tracking responses, and incorporating comments. At other companies, the Controlled Document group takes over once the document goes to cross-functional review.

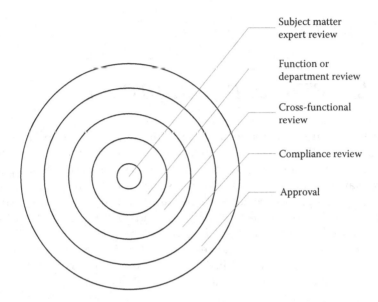

Figure 10.1 The rings of review for a cross-functional SOP. The size of the circles does not represent the number of reviewers but rather the movement of the SOP into the broader Clinical Development environment.

Expert review

The first round of review has to be a review by experts. If a working group developed the new process, they can act as expert reviewers of the SOP. Because translating a process map to text always identifies issues, they may see the process in a new light when reviewing the text. If the process has been piloted, some of the people involved in the pilot may also act as reviewers. These people familiar with the process are often called subject matter experts or SMEs. When the process is not new and the SOP has only been revised, an expert review is still warranted. Some departments keep a list of staff members who may be considered experts on a given topic to make this kind of review easier. If such a list is not available, it may be necessary to request that the management team provide the names of people available to provide expert input on an SOP. As a rule of thumb, the SOP should never proceed to the next round of review without having someone other than the author(s) look at it, even if it was a minor revision.

Function or department review

The SOP next goes out to a broader group within the function or department sponsoring the SOP. This is true even for cross-functional SOPs because it is almost always true that an SOP impacts one group more than another or that the process is owned or spearheaded by a group. (This is sometimes formalized as identifying a BPO for each SOP in Clinical Development; see Chapter 6.) The SOP is reviewed by the BPO group first to ensure it represents their interests properly before it moves to the next ring out.

Who should these department reviewers be? It may be tempting to send the document to people in the department who are experienced with the procedure in question or to those who are otherwise very interested in that particular process. However, this approach can miss critical input as it avoids the broad range of experience that will ensure the procedure will work for all types of studies. It also becomes difficult to identify a group of people for each SOP or supporting document that may come through. The best solution is to identify a review team whose members represent the variations within the group. For example, if there is a strong therapeutic area alignment at the company, then all therapeutic areas should be represented in functional review. For large organizations with multiple geographic locations, all the locations should be represented. If the functional group is large, representatives from the subgroups should be represented. There are often differences between the ways Phase I, Phase II and III, and Phase IV trials are run, so these should also be represented. Note that a single reviewer may satisfy more than one of these conditions, as might be the case for an employee that represents one site, one or two therapeutic areas, and has worked on Phase II and III trials.

The job level of the members of the review team can be anything from relatively junior staff to senior management. At companies where management is not that familiar with the work being performed, it would be better to have people who do the work review the documents. Less experienced staff members may be able to provide useful insight into how clearly the SOP is written and whether important information is missing; more experienced staff may be able to raise questions about the process or point out special circumstances that need to be considered. When the review team is composed of nonmanagement staff, there may need to be an additional level of review at companies where SOPs are required to be "management approved." Management input should also be required when significant staff resources are required for the process that do not fall into normal work assignments.

Incorporating comments from department review sometimes results in substantive changes to the process. When this occurs, a re-review may be necessary. This may involve sending the revised SOP to all of the original reviewers, only to those that provided comments (because not all reviewers review and sometimes they have no comments), or only to those reviewers who had significant input in the area of the document that was changed. Because review can be a lengthy process (see below), keeping the re-review to a targeted group may be the best choice in moving the document forward.

Cross-functional review

After all the department comments have been received and adjudicated, the SOP is ready to go out to the cross-functional reviewers. Unless the SOP is limited to a single function, the SOP *must* be reviewed and agreed to by every department whose staff may fill a role with responsibility in the SOP. (Refer to the section of Chapter 9 that describes the responsibility portion of the SOP template.) Even when a role is listed with no other purpose than review or approval of a document, those roles need to be given a chance to comment as to whether or not they agree that their involvement in reviewing or approving a document is appropriate. They may not agree to be a reviewer because the document does not impact them, or they may ask to be an approver to ensure they have a say in the final content of the document in question. It is not unusual for new SOPs or SOP revisions to be posted without input from a group who then wants to know, "Who agreed to this step? We did not!" Clear rules for cross-functional review are essential. In Example 1, we see how cross-functional review was essential in determining which resources would be committed to carrying out a step.

Example 1

In an SOP to be followed when preparing case report forms (CRFs) to be included in a regulatory submission, there was a task to compare two lists to ensure the CRFs being included reflected the final text narratives in the submission. The step involved comparing the list of subject CRFs being prepared against subjects listed in narratives. The comparison was not a simple one-to-one exercise and might require research to ensure everything was accounted for. During cross-functional review, all the groups involved in the process initially declined to perform this activity due to lack of resources in their groups. They all agreed that the activity needed to take place but no group agreed to do it. This went around and around several times, and in the end an agreement was negotiated whereby the initial comparison and research was done by one group and any discrepancies would have to be researched by two different groups depending on the source of the discrepancy. This was clearly not the most effective means of performing the task but it spread the resources required across groups. If the SOP had not been reviewed by all groups involved and a resolution negotiated, it is likely that this comparison step would have been skipped for some submissions because of lack of staff.

At some companies, it is the responsibility of the BPO to conduct cross-functional review and at others it is managed by the Controlled Document group. In either case, the sender has to know the name of an actual person or a specific email distribution list for each of the functional groups mentioned in the SOP in order to know who to send the document to. Just as a functional group may have a review team with specific individuals identified for process document review, so too should there be a list of contacts from each function who will receive SOPs from outside their group for review. This list is best kept by the Controlled Document group but sometimes maintaining the list is not in their remit. That leaves each functional group or department to figure it out. For some of the departments, a department process and training group may be the contact, for others, there will be a single person who has volunteered to perform the review. Figuring this out each time an SOP comes through can be time consuming and frustrating; the best practice is to have the Controlled Document group maintain the list and be responsible for sending the SOP for review to ensure that the document is actually sent to *all* the impacted groups for their comment.

Cross-functional review often turns up significant issues and Example 1 shows just one example: certain roles do not agree to certain responsibilities. It is also common for some groups to ask to be removed from or be added to review and approval tasks in the SOP. Cross-functional review may also identify places where yet another department should be involved. Because these issues have to be negotiated across functions and sometimes must be raised to more senior management, adjudication of the comments from this round of review can take weeks, and re-review of some kind, similar to that mentioned in functional review, may be required yet again.

Compliance review

After cross-functional review, the SOP should be pretty much final. Everyone has agreed to the process and to the role of their function in the process. But there is another round of review that should take place on this near-final draft—compliance review. In compliance review, a reviewer from the Regulatory Compliance group or an appropriately experienced member of the Controlled Document group reads through the SOP, almost as an auditor would. The compliance reviewer would check for the following:

- Activities or tasks that would appear to be contrary to or do not meet expectations for compliance with regulations or guidelines
- Places where additional evidence would be needed to show compliance
- Appropriate collection of approvals and appropriate level of approvers

At some companies, the compliance reviewer might also review the SOP in the context of other corporate SOPs for general approach and level of detail. When the compliance reviewer has been consulted during the development of an SOP or when the SOP is straightforward, there may be few comments. However, in some areas of high interest to inspectors such as medical review and account management, the compliance review may result in significant feedback that then has to be addressed. If addressing these issues results in substantive changes, the whole document may have to go back one or two rings of review.

Approver review

Because SOP approvers (see Chapter 11) should never see the SOP for the first time when it is sent to them for approval, they should receive the SOP for a courtesy review prior to the approval step. At some companies, the SOP is sent to the approvers in parallel with compliance review; at others, it is sent in parallel with cross-functional review. Approvers that are actually knowledgeable about the process probably want to see the document earlier, so that their input may be considered. This would argue for review along with cross-functional review. When approvers come from high-level management and rely on their department or function reviewers for assessing the process, then sending it in parallel with compliance review is fine. Even when comments from the approvers are not expected, they should still be sent a copy with the understanding that they could comment and have their comments incorporated prior to posting. Some vice president level approvers like to comment, others do not.

Controlled Document group review

The SOP has one more stop prior to posting, this one with the Controlled Document group. The Controlled Document group's review is not strictly a review, but rather more a processing or formatting step. Nonetheless, it may result in feedback that requires the SOP author to make changes. In this step in releasing a document, which is different from the compliance review above that may also be carried out by someone from the Controlled Document group, the Controlled Document representative will follow a checklist to ensure the document is ready for posting. One step will be to ensure the document adheres to the document template in structure and style. Another step will be to review all references and ensure the document titles and identifiers are consistent and correct. Additional checking might include checking that the document uses the *Definitions* section properly, spells out acronyms at first use, and uses company-standard role and department names. Some of these checks may involve some minor discussions with the SOP author.

Best practices for review

In healthy and working controlled document environments, SOPs and related documents should be coming through departments for review on a nearly constant basis. Some will be new, some will be revisions, and some will come in from other departments for cross-functional review. Departments that accept that document review is an ongoing effort and set up review groups for that purpose, as described above, will also benefit from implementing some additional best practices for sending documents out, tracking review, and incorporating feedback.

Preparing documents for review

When completely new SOPs or work instructions are sent out for review, it is important to send along associated documents. For example, if a new SOP mentions two forms and a template, the SOP author must draft the two forms. The template may or may not be ready at this point, but even if it is not in a final draft (as might be the case if the document was going to be a department-managed document), the author should mock up the general format of the template. When the package of documents is sent out, the author can specify whether or not the additional documents are to be reviewed or are only being provided as an example for reference.

If a document has been revised, the author must make an assessment as to whether the document can be sent to reviewers with changes tracked or if the changes made were so extensive that tracked changes would make the document confusing or difficult to read. Tracked changes are definitely the preferred method for limited changes, and the author can request that reviewers limit themselves to reviewing the changes—though a fair number will not limit themselves and will take the opportunity to go through the entire document. (The SOP author will then have to decide whether or not to incorporate those changes. One would think that busy reviewers would be grateful to be limited to tracked changes, but indeed many will take advantage of the opportunity to review more broadly!) Compliance reviewers may have specific preferences as to whether they want to see tracked changes or not; it is best to contact the reviewer(s) to find out before accepting all changes in a document before sending it out and losing the details of edits made.

Sending out for review

As noted above, the BPO or the Controlled Document group will be sending documents out for review and will be maintaining the list of contacts or email distribution lists for all departments. It is very helpful to use a standard format for the email requesting review so that reviewers can quickly identify what needs review and when it is due. Example 2 is an example of such an email. Groups may still attach documents for review to the email, but adjudicating the review is easier if the document is posted to a shared file location, where all reviewers can add their comments into the same document. This also allows them to see the comments of other reviewers, which both reduces duplication and allows reviewers to comment their agreement or disagreement with what others are saying.

Example 2

An email format that always has some consistent language but is flexible enough to accommodate different review needs helps both reviewers and the person responsible for sending the document out. Create a standard subject line first. It may read something like the following:

> *REVIEW REQUIRED – Revised CO-012-SOP "Trial Monitoring Plan" respond by Friday, October 9*

The body of the email might have a structure similar to the text below.

> *You are being sent this document for review as a member of the Clinical Operations document review group. Please provide your comments by close of business on Friday, October 9th. You may delegate this review. If you would like to review but cannot provide comments by the deadline, please notify us with a date by which you can reply. If we do not receive a response, we will assume your agreement with the contents of the document.*
> *[Attached or At this link <link>] you will find the following:*
>
> - CO-012-SOP, Trial Monitoring Plan
> - Trial Monitoring Plan Template
>
> *The SOP has been revised to ....*
> *The Trial Monitoring Plan Template reflects changes to ...*
> *If you have any questions regarding the process this document reflects, please contact Ann Author (AAuthor@Corp.com).*
>
> *Regards,*
> *Clinical Operations Process Group*

Tracking reviews and reviewers

If we accept that document review will be an ongoing activity, then it can happen that reviewers for any one of the rings of review can be overwhelmed with documents on which to provide feedback with overlapping due dates. A very useful and simple method for avoiding review overloads is to keep a review calendar that can be accessed by anyone in the department. Add the review period to the calendar as a multiday event and label each event with the kind of review (e.g., expert, functional, cross-functional, and compliance), along with the name or identifier of the document. If the calendar supports color, use different colors for the different types of review. Although it may sometimes be necessary to send out multiple document packages for review to the same group in the same period, try to avoid this even if it impacts the date on which a document can be posted.

Another tool with a great deal of value is a simple spreadsheet listing the people to whom a document was sent in each round of review. Mark in the spreadsheet who replies and who delegates to another person to easily be able to reply to the question "Who reviewed this? Did anyone from my group review this document?" Include columns for the function or department of the person, the due date for the review, and any additional comments. The comments column can be used to note cases where a reviewer provided feedback only by email or simply said, "No comments from me." Although this information could be extracted from emails received around the due date (if the emails are saved) when those questions come—and they *will* come—it is a lot faster and more convenient to get it from a spreadsheet. The spreadsheet has an additional use: if reviewers' participation in the review group is part of their ongoing work assignments and will be assessed as part of their yearly performance review, the spreadsheet can provide the metrics managers need to determine if the employees have met that particular goal.

Adjudicating feedback

The email text found in Example 2 includes an option to either attach the document review or to provide a link to a shared location where all reviewers will add comments. Having all reviewer comments in a single version of the file is a benefit to both the SOP author and the reviewers, who can then avoid repeating the same comments or chime in when questions are raised. Current best practice is to use the shared file location, but when it is necessary to use an email attachment encourage the reviewers to reply-all and add on to the comments of those who have gone first. For reviews going to a large number of people, one company required that reviewers add all of their comments to a separate spreadsheet, providing the section number and comment only in that form, as it had become unmanageable to fit all the comments in a single document or to merge a large number of documents.

This leads to a very important point: the SOP author going through comments must do so in an organized way to ensure that no comments are overlooked, even if they are declined. Reviewers like to know that their comments are not being ignored and will review better in the future if they feel they are being taken seriously. It is common that a reviewer will notice that their comment was not incorporated in the final document and ask about it, and the SOP authors need to be able to explain what the decision for that comment was. One technique to track responses or decisions for all comments is to make a copy of the document with comments at the end of the review period. The SOP author then goes to each comment and adds a response to it. Some comments may be accepted and implemented, some may need research, and some may be declined. The technique of using a spreadsheet, mentioned above, makes this assessment especially easy. Whatever technique is used should help the author track comments and make sure all are addressed and incorporated as appropriate. After all comments have been adjudicated, the SOP author should keep a copy of the result of the adjudication. Some groups post the adjudication responses at the end of the adjudication period so that reviewers can check for themselves how their comments were addressed.

Sometimes, a reviewer may not understand a particular part of the process or make a comment that in some other way warrants a longer response. Reviewers like it when the author sends a polite email to them directly explaining a particular reasoning or clarifying the process. It is not unusual for this email to a reviewer to also result in a clarification of the language or provide information on what to include in training.

A particularly challenging situation arises when too few reviewers provide feedback or a key reviewer has not weighed in. Even though the sample email text in Example 2 says that if no response is received it will be taken as acceptance of the process, SOP authors are often aware that particular issues are important and that a specific reviewer or a specific function has to agree to the proposed approach. The coordinator of the review has to make the assessment of whether to move ahead or hold up the process until the needed comment has been received. The decision of whether or not to proceed often depends on the importance of the document to the organization and this factor mixes with company culture. At some companies it is never OK to proceed without having the input of certain key contributors; at other companies the message is, "They had their chance." Whenever possible, the coordinator who sends out the review email should send a targeted message to those reviewers or groups whose input is particularly important, making this importance known and asking them to identify replacement reviewers if needed.

Approval

When all the rounds of review are complete, the SOP can go out for approval. This brings up an important question: who approves any given SOP? The main criterion is that the approver must be senior enough to speak for the function(s) involved and agree to the commitments being made in the SOP. The approver for a given department or function is generally the head of that function and so is at least a director or associate director.

At many companies, the head of each group with responsibility in the SOP will be called upon to approve the document. So if Clinical Research (e.g., medical monitors), Clinical Operations (e.g., study leads and site monitors), and the Drug Safety group are involved in the procedure, there would be three approvers. Some other companies like to keep the number of approvers to a minimum and go up the organization chart and request even more senior management to approve. For example, if Biostatistics and Clinical Data Management both report into the head of Biometrics, the head of Biometrics, often a vice president, is asked to approve an SOP that involves both groups and the heads of the two functions will *not* approve the document. Executive-level approvers often do not know enough regarding the details of the process to evaluate an SOP, and many of them will not review before approving, relying instead on the work that has been put in by the reviewers. This approach of moving up the organizational chart can be taken to an extreme as in the company where the head of Clinical Development was asked to approve an SOP that had heavy cross-functional involvement. It is unlikely that the head of the entire development group would know enough about the procedures in question to judge whether or not the SOP is appropriate, so the value of such high-level approval becomes questionable. In companies that follow this policy, the heads of each of the individual departments should be given an opportunity to review even when they will not be approving, as they are the ones responsible for committing resources. As noted above, all approvers, whether they typically comment or not, should be given the opportunity to review an SOP before receiving it for approval.

In addition to approvers who represent each of the impacted functions, some companies also have a representative from Regulatory Compliance approve each SOP. This can be valuable to show that Regulatory Compliance review of the document has taken place and no issues of compliance to regulations or good clinical practice have been identified. Their approval indicates that the Regulatory group stands behind the decisions made regarding the process that are reflected in the SOP.

Activity	Allotted Time (business days)
Expert review	1 week
Adjudication	3 days
Functional review	1 week
Adjudication	3 days
Cross-functional review	1 week
Adjudication	3 days
Compliance and approver courtesy review	2 weeks
Adjudication	3 days
Controlled Document group processing	1 week
Obtain approver signatures	1 week
From draft to posting for training	9½ weeks

Figure 10.2 Time estimate for a new or heavily revised cross-functional SOP from a review-ready draft to its posting for training.

Because the approvers are typically senior management staff, getting the approval taken care of usually adds at least a week to the SOP development time, as shown in Figure 10.2. Even though many companies have moved to using validated electronic signature software, they still have to allow time to get the attention of the approvers. The Controlled Document group is usually in charge of getting in touch with the approvers of an SOP and training them, if necessary, on how to log in and apply their electronic signatures to the document.

How long does this take?

Factoring in each of these rings of review is what makes it hard to update an SOP quickly. Although it might be possible to do a SME review and ask for comments in a couple of days, a week's turnaround is tight for most reviews. Cross-functional reviews often require two weeks or more but could be done in a week for urgent documents. After each round of review, the SOP author and possibly other SMEs have to go through all of the comments and determine the right response—this could take just one day or it could take a couple of weeks if the review turns up critical issues with the process. Figure 10.2 shows a realistic time period that is probably close to the minimum time for a cross-functional SOP. This tight schedule might work if there are no major holidays in the way, if the compliance reviewer is not out of the office, and if the SOP author and SMEs are not bogged down doing other work. It also assumes that there is no need for significant rewriting or re-review after any of the reviews. Given these assumptions, the time estimates shown in Figure 10.2 could be used to *start* a project plan, but it would be unrealistic to expect to complete the job in the nine and a half weeks shown unless there were a serious business need to post something very quickly.

Reviewing documents from other departments

There are a few differences for reviewing documents sent from other BPOs compared to reviewing documents authored in the function. When there is a single contact to whom review requests are sent, it does not mean that only that one person should review the document. Ideally, the entire review team for the function reviews the document and returns comments to the function contact. (See Example 3 for a situation where limited cross-functional review led to a serious problem with a posted SOP.) The function contact should assess and consolidate comments to return a single response from the function as a whole. In the best case, the function contact will work within their own function to resolve differences of opinion before sending it back to the BPO or Controlled Document group. There may be some cases where the function contact determines that changes to an existing SOP do not significantly change the activities already being conducted and so limits the review; this should be the exception as broader review will generally lead to a better outcome.

Example 3

The Clinical Coding group at a company decided to change the process for coding reviews in two ways. They decided that the coding staff would communicate coding issues to the data manager and medical monitor rather than having the data manager coordinate the process. They also wanted to add a regular cumulative coding review by the medical monitor in addition to the reviews for purposes of analysis or study database lock. The cumulative review was to include all terms coded to that point, so that for large studies there was a very large volume of terms, some of which would be reviewed over and over again. Unfortunately, the medical monitors and the clinical data managers did not agree with these new processes. The data managers felt that they were a better lead for coordinating final coding approval and they did not see that the cumulative review added any value above the normal coding review already taking place. Medical monitors did not want to be involved too early in coding issues and did not agree to the extra work that the additional cumulative review would introduce.

This SOP revision had not gone through the full Clinical Data Management review team; a single person had reviewed it.

The medical monitors may or may not have had anyone review the document. In any case, review was complicated by it being a very long SOP with 30 pages of process in text outline format where "medical monitor review" was just one small step in huge document and the new version was not sent around with tracked changes. These issues only became clear when the Coding group presented a coding overview to the Clinical Data Management department with slides showing this new process. After questions were raised, the Coding group made the cumulative review optional, which was possible because the SOP language was somewhat vague. The Coding group then revised the SOP to return coding review coordination back to the data manager.

SOP of SOPs

The SOP of SOPs should explicitly list the kinds of review that are expected. The details of how to carry out expert and functional reviews can be left to the functions, but the SOP of SOPs should clearly stipulate that cross-functional review must include any function or department with a responsibility in the document, including those that have review and approval activities only. It should also be clear who (the BPO or the Controlled Document group) is responsible for managing the list of cross-functional review contacts and for managing that review cycle.

The SOP of SOPs should also make clear what kind of approval is needed for each kind of controlled document. In general, a high-level department manager, generally the head of the department or function, must approve SOPs. The example SOP of SOPs in Appendix 2 uses this approach rather than that of moving higher in the organization. The example also explicitly includes a review for approvers.

Although the SOP on SOPs would not include information about how long reviews will take, the Controlled Document group would provide SOP authors a great service by providing a workflow diagram and some estimates of minimum time required for each type of review, similar to those listed in Figure 10.2.

chapter eleven

Posting
Setting up for success

After putting a great deal of effort into developing or revising a document, most SOP authors and even many Controlled Document groups miss some key activities around posting that are needed to ensure that everyone who needs to does indeed know about the new process and can follow it on the day it becomes effective. Careful planning and communications can set the organization up for success in making use of new efficiencies and in compliance.

Is it really ready to post?

When an SOP or work instruction is posted as effective, all the documents in the hierarchy that directly support it (see Chapter 2, Figure 2.2) have to be ready also. It is particularly important that all supporting documents and training have been updated to be aligned with the changed or newly released procedure. It is best to figure out the package of related documents at the *start* of any revision but in any case, before posting for training, ensure that the following have been reviewed and revised as necessary to ensure they are aligned with or reflect the changes:

- Documents, both controlled and department-managed, that reference the central document
- Forms and templates directly referenced by the central document
- For SOPs, all related work instructions that support the SOP
- Existing training materials
- Process diagrams maintained outside of controlled documents
- Any timelines or project plan templates that are tailored to the process

In some cases it may be necessary to time the retirement of an SOP with the release of a new document or to time the release of a department-managed document with the retirement of a controlled document. The former is especially important because there should never be a case of conflicting SOPs both being effective at the same time. If many documents are impacted, it may not be possible to have them all updated prior to, or simultaneously with, the posting of the central document, but the information regarding which document updates will follow later can be an important part of the communications around posting, as discussed below.

Communication plans

When an SOP posts for training, it triggers the most important elements of a communication plan, but communications may have begun earlier for new SOPs or greatly changed procedures. An automatic notification from the controlled document system that training is required for a new document should *never* be the only information an employee receives! A good assessment of what kinds of communication are needed is an essential part of a change management.

Preparation

Any of the methods of communication should be able to address the following:

- Related and/or impacted documents
- Training requirements
- Cut-over rules

The list of related and/or impacted documents will have been identified as part of the revision or during the assessment of whether or not the document was truly ready to post (above). If they are not all being released as a package, prepare timelines and provide estimated dates for any lingering updates.

For training, employees will want to know what kinds of training they will be required to take, when it will be available, and when it must be completed. In particular, they will want to know if they will need to do anything beyond a read-and-acknowledge for the document. If new training is required and it is instructor-led, when will it be available? For those that have not yet taken a particular training, they will want to know if it has already been updated to reflect the new or changed procedures. For example, if a new electronic data capture system will be used for all new studies, it may be that site monitors currently working on other studies do not have to complete a new instructor-led training right away (though they would have to read the SOP) and can wait until they work on a new study using the new system. However, that time period may be too undefined and the company may choose to say, "Complete training before working on a study using the new system or before the end of year, whichever comes first."

Cut-over rules explain how the impacted departments or functions will move over to using the new process and should include all possible circumstances for studies. Cut-over rules are discussed in more detail below.

Communication plan options

For *both* controlled documents and department-managed documents, consider the types of communications described below and select the ones most appropriate to the change. The bigger the change, the more types of communications should be used. Except for very simple document revisions, it is usually best to combine approaches so that it is never only email or only live presentations.

- **Functional and cross-functional review**
  Reviews are a kind of early communication because key members of the functional groups impacted by the SOP will see early versions of the document. Ideally, they will solicit input from other members of their groups and perhaps discuss the new document or process in their groups. A good cover email for the review that provides background to the process or document changes can make for an effective early communication.
- **Coming-soon presentations**
  For important changes or new processes, it can be very helpful to give brief presentations well in advance of the initial SOP posting to give impacted groups advance notice of important changes coming. If other groups will need to adjust timelines, training, or staff resources, this early notice is essential. A business process owner (BPO) representative or members of the working group for the process can request to appear at functional or department management meetings, and functional or department all-hands meetings. These visits should occur after completion of cross-functional review because it often introduces substantive changes to responsibilities. Changes can still be introduced during compliance review, but, because meetings such as "all-hands" meetings may not take place very often, it may be necessary to stick to a very high-level discussion and provide a very rough timeline for document release in the presentation. Coming-soon presentations are fairly rare.
- **Email at posting for training**
  When a controlled document posts for training, it is likely that the controlled document system will automatically send out a notice to every person associated with a role to whom the document has been assigned—but there won't be such an email for

department-managed documents. Because the automatic notices are not especially helpful, an informative email at the time of initial posting, such as that in Example 1, is a requirement for all kinds of documents. Note that cut-over rules may be included in this email or may wait until the document becomes effective.

- **Informational presentations**
  These presentations are similar to the coming-soon presentations but have the final version of document available and the information around document release, including cut-over rules. When kept short and relevant to the audience, informational presentations are very valuable because busy people often don't read their email and this is another way to get important information to those people who need it. These presentations can be given at any combination of department, function, or subfunction meetings. In companies that are organized by therapeutic area (TA), TA group meetings within a function may be appropriate. As in the coming-soon meetings, the BPO or representatives from a working group may present to different departments—and the presentations may be different, so that each group hears the information that most pertains to their work activities. Presenters should assume, again, that this message will also not reach everyone and should make the presentation slides or other materials generally available after the in-person presentation.

- **Email on effective date**
  A second email when the document becomes effective should also be required for all documents where the dates of posting and becoming effective differ, such as SOPs and work instructions. The main message of this email is that the documents now apply to all studies according to the cut-over rules that have been defined. The format shown in Example 1 serves well here also.

- **Training announcements**
  If special delta training (see Chapter 15) has been created or if new training courses are available, there may be additional announcements when those are scheduled.

Example 1

A structured email format used for the posting of all documents helps provide the critical information and allows readers to focus on items most important to themselves. The sample text in this example might apply to the release of an SOP for training that had updates to the reviewers and approvers of a study document. Slight modifications would tailor this message for use when the same SOP becomes effective. This kind of format can be adapted for notification of posting for all kinds of documents, not just SOPs. Because not everyone reads their emails, departments can keep a history of these announcements on their internal web page.

FFF-001-SOP "Title of the Document" Posted for Training *or* Effective Today	
Summary	The process for drafting "Doc Name" has been changed to improve efficiency. Approval by the [ROLE] will no longer be required.
Action required	Begin drafting any new "Doc Name" using the new template if it will be approved on or after [dd-mo-yyyy] when the SOP becomes effective.
Cut-over rules	• All new studies that have not yet drafted a "Doc Name" must use the new template. • Studies that have already begun a draft of "Doc Name" are not required to switch over to the new template but must use the new approval page once the SOP becomes effective. • When revising "Doc Name" after the SOP effective date, replace the approval page with the new one aligned with the SOP.
Related documents	• Form: FFF-001-FRM-1 "Name of Form" has not been changed. • Department-managed Template: "Template Name" has been updated to align with the SOP revisions, including a new approval page.
Training	Read and acknowledge the SOP in the CDOC system now. Instructor-led training on the "Name" process has been updated to reflect the SOP changes.
Document location	FFF-001-SOP: in CDOC *here* FFF-001-FRM-1: in CDOC *here* "Template Name": at the department-managed location *here*
For questions	Email: Doc.Questions@Corp.com

SOP release cut-over rules

The cut-over rules are the most important item to focus on at the time of an SOP release, because this is the area that can lead to confusion for the staff and, in the worst cases, the appearance of noncompliance. If a new SOP for a new procedure is introduced, it is not always obvious if it applies only to new studies starting *after* the SOP becomes effective or if it applies to *all* studies in conduct because it can be implemented retroactively. Even in cases where we say the new procedure applies only to new studies, we have to define what *new* means quite specifically. Some examples will illustrate the point and demonstrate how implementation memos can address the problem.

Why do we need rules?

The example of implementing a single study plan that documents all the different kinds of data quality procedures and data review that many companies have been moving toward as part of changes to monitoring illustrates the reasons we need to be clear about how a new SOP is to be applied. These data review plans may list highlights of medical safety review, study team manual review of listings, and refer out to the study monitoring plan and the study edit checks. This kind of plan has a variety of names, including "study data review plan," "clinical data review plan," or "study data quality plan," and is associated with a template for the plan. When the SOP governing such a plan is first released, we know it will apply to new studies. But does *new* mean the protocol has not yet been approved, that the study database has not been built, that there is no subject data yet, or that it has only been accruing data for three months out of an expected three years? Next we ask whether this SOP should apply to existing studies. An auditor reviewing this SOP a year from its release might assume that all studies, not just new studies at its release, are required to have a data review plan, since it is not a computer system or application dependent procedure, and because the scope did not say otherwise. Example 2 shows how one company decided to implement data review plans.

Example 2

When a medium-size company introduced the data review plan concept, it opted to make it retrospective; because all ongoing studies should already be doing data review anyway, they reasoned it would not be difficult to document those efforts and would have value to the teams of longer running studies. They defined parameters for studies in all stages. To define *new* studies, this company chose a specific date and criterion. The cut-over rule for new studies was as follows: "All studies that do not have any subject data in the study database on the effective date for the SOP are considered new and must follow the provisions in the SOP for approving a study data quality plan."

For ongoing studies, the company felt that all studies should indeed create a data review plan and follow the procedure in the SOP, documenting what they are already doing. But fast running studies (e.g., Phase I) that were already in conduct would not be required to switch over. The SOP business process owner, after consulting with impacted functions, decided to set a six-month period for creating a data review plan starting from the January effective date. They came up with the following cut-over rule for existing studies: "Studies that have subject data in the study database on the effective date for the SOP and will have study lock on or before June 30th will not be required to create a study data quality plan. All studies continuing beyond June 30th will be required to approve a data review plan on or before that date." Notice the use of "approve" rather than "create" to indicate that the plan is expected to be final on that date and not just getting going.

Another kind of cut-over rule or instruction must be defined when the reviewers or approvers of a particular study document change. The new approvers apply on the date the SOP becomes effective, and that is clearly understood by all and easily accommodated in new documents by providing an updated approval form or page in the document template. The matter is a little more complicated if the document or plan is already in use for studies and was previously approved by a different set of roles. Even if the study document already exists, the revised SOP procedures apply. Although we would not go back and require reapproval for the existing documents with the new set of approvers, when the study document is *revised* as part of its natural course, the new approvers will apply. Because the person responsible for the study document revision will take the electronic version of the previously approved document version to make changes, they will have the *previous* approvers in that document (unless a standalone form is used). The study document author must know to replace the previous approval page with the new one or edit the approvers to reflect the new requirements. This situation can be confusing, so creating clear cut-over rules, and even a table, may be necessary. Example 3 continues with a data review plan, but this time assumes the SOP already exists and has just been revised to have a different set of approvers.

Example 3

Consider the case where a version of the SOP governing data review plans is already in effect and the current version requires approval of the plan by the study leader, the medical monitor, the biostatistican, the clinical data manager, and the statistical programmer. On revising the SOP, the groups all agree that the statistical programmer should review the document but does not need to approve it. If it take three weeks on average for the plan to be drafted and reviewed prior to approval and the training period prior to an SOP becoming effective is two weeks, then we might have a complex set of cut-over rules as follows:

- The new data review plan template, with the *new* approvers on the approval page, will be posted when the SOP is posted for training to permit study teams to draft a plan using the latest version. If the plan is ready for approval prior to the date the SOP is effective, the study document author should use the *previous* approval page.
- Once the SOP is effective, the new approvers apply and the approval page must be used. The study document author must copy in the *new* approval page when revising a previously approved version or one drafted using the previous template version.

Implementation memos

If the cut-over rules are documented only in the release emails and presentations outlined above, we are back to a situation where there may be an appearance of noncompliance as would be the case where a new procedure does *not* necessarily apply to existing studies. Ideally, the Controlled Document group and the controlled document system can formalize the cut-over rules into an implementation (or release) memo that remains associated with the SOP version and is available for inspections. If the software used to manage controlled documents does not have this functionality built-in, workarounds may be possible. Options include the following:

- For procedures that apply going forward only, add this to the scope of the SOP.
- Use another document type, such as a work instruction, associated with an SOP, to formalize the cut-over instructions.
- Store implementation memos separately (perhaps as a department-managed document). In those cases, care must be taken that access is controlled and all older versions are retained.

Posting SOPs

When all approvers have indicated they are in agreement with the SOP, the SOP is ready to be posted for training but it is not yet effective. This is a very important distinction that staff members need to understand: when an SOP posts for training, you do not follow it yet. Continue to follow any existing SOPs until they are replaced or retired when the new SOP or version becomes effective. This period, sometimes called "prerelease," is a period of time during which everyone impacted by the SOP can receive training on the new procedure—even if that just means everyone has to read it. The length of time allocated for training at a mid-size company is a minimum of two weeks, if the only training required is to read the document. A longer period of time may be required if additional computer-based or instructor-led training is needed, as it might be for a complex new process associated with a computer system. (See Chapter 15 for further discussion on training for SOPs.)

Making the new document or version available for training while still keeping the currently effective document available for it to be followed can be a stretch for some controlled document systems. As we saw in Chapter 2, making sure employees can always access the right document, the one in effect, is a critical attribute of quality systems. Some controlled document systems address this issue by making the new SOP or version available only through the training interface; it is not generally available in the document library. This causes difficulty when users want to compare the old and new or look at the new version again after completing training but before it becomes effective. Other systems will have both old and new documents available in the document library identified only by their status of "prerelease" or "effective." This difference can be subtle and can be confusing to users; including a watermark can help somewhat. Ideally, both documents would be readily available but easily differentiated.

When the training period is over, the SOP becomes effective. Many controlled document systems do not send out notification of this change of status to anyone but the document coordinator in the Controlled Document group. The very important email directing readers to begin following the new SOP must therefore come from the BPO contact, who should send out an email notifying all impacted groups that the SOP is now effective and include any updated information on related documents and training.

SOP of SOPs

Although it is rarely found in current SOPs of SOPs today, the requirement to have the BPO provide an implementation memo would improve SOP roll-out. More typically, the Controlled Document group will make the option of a memo available if the SOP author brings up complex issues of implementation. The SOP of SOPs must differentiate posting for training and posting as effective.

Maintaining SOPs: Maintaining compliance

chapter twelve

Deviations from controlled procedures

Deviations from, or changes to, procedures found in controlled documents will happen almost as soon as the document is released. In some of these cases, the study team may not even realize that they have not followed a particular procedure correctly—especially if the outcome of the procedure was not adversely affected. At some point, often at the time that the study is closing down and review of the trial master file (TMF) is taking place, someone will notice. We will call those situations *retrospective deviations*. In other cases, a study team *plans* to follow a different procedure—whether to look for new efficiencies, adjust to changes in regulation, or pilot a new system. We will call these situations *prospective deviations*. In this chapter, we will look at the two kinds of deviations, how they are recorded, and what should be done with them once they are identified.

Retrospective deviations

Retrospective deviations are unplanned deviations from controlled documents, identified after they have taken place. Although they can be uncovered in all kinds of ways, it is most common to realize that an SOP has not been followed when reviewing documents in the TMF, often at the time of study closeout or during preparations for an expected internal or external audit. An employee collecting documents listed on a TMF checklist realizes that a document is missing completely or missing certain approvals and begins to investigate. Another common way deviations are discovered is during study conduct when handing over responsibilities from one person to another. The new person, coming in with fresh eyes or with experience from another therapeutic area, might say, "Wasn't this was supposed to be done differently?" Even when a deviation is identified during study conduct, it is most often too late to do anything about the past process, though if the deviation is found in mid-study the activity could still be taking place, in which case, it can be changed the next time it is followed.

Whenever and however a retrospective deviation is identified, the first step is to determine whether the deviation introduced any risk to the subjects in the trial, to data quality, or to data integrity. One might argue that *any* deviation from an SOP is a risk to the trial, but practically speaking that is probably not the case. A document missing one of four signatures is likely still fine, especially if the review process took place but the approval activity missed a step. However, if there is a chance that medical review of a particular group of subjects was not performed or if data was unblinded prematurely, there is a concern about the study and a more detailed review and evaluation of what happened should take place immediately. For the more serious deviations, the process of figuring out what risk the deviation introduced may also lead to identification of actions that can mitigate that risk.

Analysis of deviations should also go to the next level and attempt to identify why the deviation occurred. There is a set of formal techniques that can be applied here called *root cause analysis*. Because typical Clinical Development staff will not have been trained in these techniques, it is very helpful if someone from the Controlled Document group or Regulatory Compliance group familiar with root cause analysis can assist in getting to the underlying problem that caused the deviation, not just identifying the immediate cause. The analysis must also determine if other studies are likely to have made the same error.

Once the deeper cause is identified, a recommendation can be made as to how to prevent this mistake from occurring in the future. All retrospective deviations should directly address this question, whether or not the form for registering the deviation template asks it explicitly. The possible range of actions is of course very large, but common approaches to preventing similar deviations in the future include improving training for staff and updating SOPs so they more accurately reflect what is being done. Example 1 demonstrates how an informal root cause analysis can help prevent future deviations when management acts on the cause. Because the analysis associated with retrospective deviations is similar to that used to address audit or inspection findings, some companies call the documentation of a retrospective deviation a *CAPA* (corrective and preventive action).

Example 1

A review of the TMF at study close identifies that the programmer responsible for filing validation materials associated with the listing reports used for medical review does not have them. The analysis started along these lines:

- Are the materials misplaced, disposed of, or were they not done?
  - It turns out that the programmer did not perform the validation steps.

- Why did the programmer not perform the validation steps?
  - Did the programmer know it was required? Yes.
  - Did the programmer know how to carry out those steps? Yes.
  - Did the programmer have time and resources to perform the steps? No.
- The programmer explains that there were not enough people available in the department to perform the required validation on all the reports being developed.

Further analysis then proceeded to ask why resources were not available and what could be done (within the reality of group budgets) to address this in the future. At this company, they would not be able to solve the resource problem quickly. That meant that it was not possible to go back and do retrospective validation for this study or other studies where validation might have been skipped. They had to assess how much risk lack of validation for listing reports introduced. Looking to the future, they had to consider whether reducing validation needs for certain report or programs using a risk analysis was an option and whether moving toward more standard reports that only have to be validated once could be accomplished quickly. All the options they considered realistic would involve updating the SOP and might require a prospective deviation (see below) in the interim. At other companies, the executive decision might have been to immediately remedy the resource situation by bringing in additional contract programmers.

Once all the information needed to document the deviation has been gathered, it is written up using the form or template associated with the SOP governing deviations. The deviation will then need to be reviewed and approved as discussed later in this chapter.

Prospective deviations

Prospective deviations are used for circumstances where a need is identified ahead of time to perform a procedure differently from the way it is documented in one or more SOPs. Requests for prospective deviations arise from the following circumstances:

- Individual study teams face unusual circumstances, such as hold-up of a final protocol, which would otherwise trigger downstream activities.
- Groups of related studies would like to take advantage of similarities rather than follow the more general process for each.
- A team or working group wants to pilot a new procedures.

- A function in the organization is reassigning responsibilities or renaming job descriptions, often as part of a reorganization.
- A business process owner needs time to update SOPs, either to reflect a process change that has already occurred spontaneously, or to complete updates of related SOPs after revising another SOP.

In the case of prospective deviations, the first step is usually to identify exactly what procedure will be followed and what sections of existing SOPs are impacted. Notice that the new procedure is determined *first* and then the impact on SOPs follows. It is not uncommon for conscientious teams who are piloting a new process to try and do it the other way around—but the deviation can only be accurate after the proposed process is clear. In describing the alternate process that will be followed, any risks should be identified to patient safety and data integrity. Ideally any risk should be the same or less in the new process than in the existing process; when any increased risk is possible, risk mitigation plans should be included.

A key feature in the request for a prospective deviation is documenting how everyone working on an impacted study will be made aware of the deviation and the alternate process to follow. This planning is especially necessary if the prospective deviation impacts all users or a large class of users of an SOP, as would be the case when a process is being changed and applied to studies before all the impacted SOPs can be updated (see also Chapters 11 and 13). If the deviation applies to a group of studies piloting a new system or process, the prospective deviation can require that the new process be taught to everyone participating in the pilots and training records can be filed with the studies, including the slides or process flows used. This approach works best when the impacted users can be clearly identified, such as the study teams for a specific list of studies. Another approach to ensuring everyone knows about a deviation is to require a read-and-acknowledge type training for the deviation document similar to that used for SOPs. This approach works especially well when the change is general and it is hard to know exactly who will need to be aware of the change. For example, when an SOP is updated but it is referenced by several other SOPs, it may not be possible to update all the other SOPs at the same time. Filing a prospective deviation explaining the impact of the SOP update and assigning the deviation to all those who trained on the impacted SOPs provides a short-term solution to bridge the time until all the other SOPs can be aligned with the change.

Many, but not all, companies impose time limits on prospective deviations. For prospective deviations, it makes sense to expect the requestor to say how long an exception to an effective SOP is to be acceptable before requiring a more appropriate action—such as updating the SOP.

Timeframes for prospective deviations to allow piloting a new process must be sure to allow for multiple studies to complete the activities in question, keeping in mind how frequently study timelines change. After the pilots conclude, all studies should be following the old process until the relevant SOPs have been updated. But, as we see in Example 2, it can be hard to convince staff to follow the old ways when pilots are successful.

Example 2

At one company, the pilots of a new process were so successful that the head of the department requested a prospective deviation for *all* studies starting up so they could all benefit from the efficiencies immediately. That broad deviation would be in place until the SOPs for the process could all be updated. The only reason this second, global deviation was permitted was because the group in question was the early development group at the company, focusing on Phase I and early Phase II studies. The number of employees was fairly small and speedy training for all of them could be easily assured. The later-phase study teams at this company had to wait until the documents and formal training were updated.

Deviations for special exceptions, limited to a single study or small group of studies, are self-limiting and the time limits are usually not a problem. A bigger problem is that of deviations that should not really be happening: those that are filed to allow for updates to SOPs to take place. Unfortunately, in the real world, the time it takes to update an SOP with even a minor change can be long and the resources available for the update may be limited, so that a deviation may well be necessary. The idea is to try and force the updates to be made in a timely way.

Whenever assessing how long to request for the deviation, remember that time passes very quickly and it is best to be generous or to be prepared for extending or resubmitting the deviation, which may require or impose retraining requirements. In general, the longer a deviation is in effect, the more likely it is that people will miss finding out about it or will forget that the deviation is in effect, leading to noncompliance to a deviation rather than to an SOP!

Documenting deviations

The storage of and access to deviations has not yet reached a state of best practice and it can be hard to find deviations, impacted SOPs, and impacted studies. Both retrospective and prospective deviations are typically filed in the controlled document system. A deviation of any kind may or may not be linked in the system with an SOP, and even if it is linked, it will only be associated with a single SOP. Deviations, especially prospective deviations may include impacts to multiple SOPs. And while the deviation may contain the list of impacted studies in the contents, the deviation is typically not indexed to the study. So, at most companies, you may be able to view an SOP and see if any deviations have been filed against it, but only if it was the primary impacted SOP. It is unlikely that you will be able to easily find if a given study has filed deviations to any SOPs. There is not even a place in the Drug Information Association TMF Reference Model (see Chapter 5) for process deviations, only for protocol deviations.

Not much can be done in this regard by a study team for retrospective deviations, but when the deviation is prospective there are some options. Accepting the idea that it would be useful during a future inspection to know if any deviations had been filed for a study being inspected, where would this information go? Many people in Clinical Development immediately assume it should be located in a note-to-file (for which there is always a spot in the TMF). That should be the option of last resort because there is no obvious organization in notes-to-file, nor would everyone even think to look there. It is much more valuable to put the information about a deviation closer to the activity. For example, a trial monitoring prospective deviation should be referenced by and filed with the trial monitoring plan. Similarly, for data management deviations, the information on the deviation goes into the data management plan and, in some cases, deviations in Biostatistics' processes can go into the statistical analysis plan. But clearly there are cases where there is no obvious study document in which to mention the deviation. For example, if the deviation is in the *process* of developing a protocol, where would that be documented? Mentioning it in the protocol itself would not be appropriate, so perhaps the note-to-file then becomes the best choice. Deviations for pilots (see Chapter 7) of complete process overhauls are not just documented *in* study documents, they are in part documented *by* completely new study documents as shown in Example 3.

Example 3

A company was moving from a specification-based model of building the study electronic data capture application to an approach that was more of a prototype method, where the study database is built and then reviewed extensively before the design is approved and moved to testing. This move was a significant change not only to the process but also to all of the documents used in the study build process. The specification of the electronic case report form was completely gone, replaced by an approval of system document showing what was actually built. An inspector looking at the study build process would request the documents that the SOP-in-effect requires and they would not be there. To ensure that there would be no confusion on what process the study followed, copies of the approved deviations were filed in the *study build* section of the TMF and the new process diagrams used to guide the team were also filed. Since this was a data management led activity, the study data managers were also instructed to include the mention of the deviations and alternate process in the clinical data management plan. All of the study database documents created as part of the process were filed in the TMF in the appropriate locations as directed by pilot documents.

Deviations from supporting documents

At many companies, the term *deviation* is applied only to SOPs—not to work instructions or department-managed documents. Even though work instructions and department-managed procedural documents are expected to be followed, there is no way to record when they are not followed nor is there a way to request permission to follow a different process prospectively. Because work instructions are most typically limited to a single function, a function or department can decide on its own to put a system into place to record deviations to both work instructions and department-managed documents. Although this introduces extra work and does not get recorded in the controlled document system, the department will still have received useful feedback on issues that might result in an update to the document in question. Recording these deviations on a department level also helps to instill the expectation of compliance to all procedural documents, whether or not they are controlled documents.

Is it really a deviation?

Inspectors and auditors are human and have their own opinion on matters of compliance; there is definitely a range on how strictly they interpret what is considered compliant and what is not. The same is true for the staff responsible for reviewing requests for deviations—whether that is someone in the Controlled Document group or in a good clinical practice (GCP) review group as recommended below. Where one reviewer may say that a deviation is called for, another may say that if the spirit of the SOP is followed, then some minor variations can be permitted. One such example is found in Chapter 9 under the section "Scope"; Example 4 provides another.

Example 4

A company was using an early electronic data capture system that required that the sponsor provide the sites with a computer

preloaded and configured with the necessary software. At the end of
each study, these computers were returned to the sponsor. The SOP
governing this process called for Data Management to ensure receipt
of the computer, but in some studies Clinical Operations took over
this step. A data manager requested a deviation for studies where
Clinical Operations performed this task. The company's GCP compliance committee decided that this situation did not require a deviation
because one functional group or the other was reliably carrying out
the step. The step was not being missed and the result was the same.
The SOP was due for a revision soon in any case, so a change request
for the step in question was submitted but deviations were not
filed. A different review committee might have come to a different
conclusion.

Because the opinions and experience of staff in the Controlled
Document group will vary, some companies have put into place a review
group for compliance questions; this group is composed of not just the
Controlled Document or Regulatory staff but includes representatives
from the key functional groups in Clinical Development such as Clinical
Operations, Data Management, and Drug Safety. The purpose of this team
is to review compliance questions regarding how to revise, correctly follow, or file deviations to controlled documents. A GCP compliance committee provides guidance on the specific processes from their knowledge
of the responsibilities of that function, together with their knowledge of
applicable regulations. If the committee decides that a deviation is warranted or if a deviation is submitted because the need is clear, they review
the details of the submitted deviation. They can comment on the proposed alternate action for prospective deviations, the risk to data integrity
or patient safety, or the planned method of communicating the deviation.
Members of the committee often have very practical input on time limits
that should be applied. Although the members of this group may not be
professionals in compliance, they bring to compliance questions specific
knowledge of their respective functional area responsibilities and common industry practices.

At companies that have GCP compliance committees, staff from any
of the Clinical Development functional groups can send deviations or
questions of compliance to the group via an email distribution list. A designated gatekeeper, usually someone in the Controlled Document group
or Regulatory Compliance, logs the requests or questions and responds
to the requestor. Members of the committee review the question or deviation and consult with each other, in the email thread or directly via other
communication, and come to a conclusion or recommendation or request
additional information. The gatekeeper ensures a response in a timely
manner and keeps records of the discussion and decision (which can be

done very easily if discussion is mostly in an email record). The value of input from members of the organization expert in their functional areas and also interested in questions of compliance cannot be emphasized enough. They can readily identify when cross-functional input is needed, when a proposed action is impractical, or explain to each other nuances of a particular activity. This additional review of deviation requests adds a great deal of value and should be implemented at all companies— even if at small companies it is just one or two functional area managers knowledgeable about the different responsibilities involved in conducting clinical trials.

SOP of SOPs

The topic of retrospective and prospective deviations is quite complex because it must take into account questions of training, time limits, compliance review, and documentation. Because of this complexity, deviations warrant their own SOP. The SOP of SOPs should explicitly refer to the SOP on deviations to controlled documents to ensure that users can find the process information quickly. Just as everyone in Clinical Development should train on the SOP of SOPs, so too should everyone train on the SOP on deviations.

chapter thirteen

Active SOP maintenance

It seems that as soon as an SOP becomes effective and starts to receive widespread attention and use, possible changes are identified. Most of the early recommendations for changes have to do with clarity of the wording or process and, if implemented, would make the SOP more clear. Sometimes, if some functional groups did not perform as thorough a review of the draft process and associated SOP as perhaps they should have, they may find an issue that they would like to have changed right away. But after that initial period the *environment* the SOP is released into begins to change and updates to the procedure are needed to reflect process changes, system changes, organizational changes, regulatory changes, changes to other documents, and so forth. This chapter aims to change the view of SOPs as documents that can remain static and untouched for long periods of time, to recognizing that they must be actively maintained. If we promote an active culture of process improvements and updates, while documenting those improvements appropriately as prospective deviations and SOP revisions, we improve both compliance and efficiency.

The environment changes

We operate in a regulated environment where regulations and regulatory expectations change, and we operate in a business environment with all the pressure to move new therapies and devices to market faster and more efficiently. Because the environment is changing, our procedures are changing and the SOPs that govern our procedures must change also. It is essential to view SOPs as documentation of best practices to be used consistently for regulated activities—but not as static practices that will not change for years.

Regulatory thinking changes

Actual regulations may not change all that frequently—take for example *ICH E6 GCP** itself, which was adopted in the European Union as a regulation in May of 1990, or the relatively recent FDA regulation

---

* International Conference on Harmonisation, *Guideline for Good Clinical Practice*, E6.

21 CFR Part 11, which was put into effect in 1997, neither of which has changed. But regulatory expectations change as evidenced by the guidance documents the FDA releases or the reflection papers released by the European Medicines Agency (EMA). Although guidance documents and reflection papers are not laws that must be followed, they do indicate current regulatory agency thinking and should be given serious consideration by sponsors and evaluated to assess whether any company practices should be updated. When a sponsor chooses not to follow a recommendation in a guidance or reflection paper, the actual approach being followed should be reviewed to determine whether it protects the patient safety and data integrity as well as the approach advocated by regulatory agencies.

Consider as an example the FDA's *Guidance for Industry: Electronic Source Data in Clinical Investigations*, finalized in September of 2013. This guidance, and the webinars presented by the FDA at its release, state that investigators should approve data that will be used in submissions to the agency (Section III.B.1.a of the guidance). As more companies are submitting data from interim analyses to the FDA, the need to have investigators review and approve that data before study closeout will be raised. When companies decide to accept this guidance, SOPs will have to be modified to ensure the investigator signatures are collected during the course of the study when the interim data will be used, instead of collecting them only at the end of the study as is now the common practice.

Another example of guidance documents impacting practice is the EMA's document *Reflection paper on risk-based quality management in clinical trials*, which also became final in September of 2013. That document and a similar guidance from the FDA are leading to changes in the way that site monitoring is being carried out. Although a change to site monitoring is not required by these regulatory opinion papers, it opens the door to reductions in site monitoring that can reduce costs while maintaining data quality—both of which are very desirable to the industry. Site monitoring practices, the associated SOPs, and all the supporting templates and documents will have to be updated to put the new approaches into use. (The data review plan examples found in Chapter 11 were spurred in part by these regulatory guidances.)

The message here is that the Regulatory Compliance group or the functional groups themselves will need to be current with regulatory agency thinking. Keeping current will help them to maintain SOPs to both reflect expected requirements (such as that for investigator signatures) and to make use of new efficiencies (such as changes to site monitoring practices) that updated regulatory thinking makes available.

Systems and software updates

The biopharmaceutical industry relies on all kinds of systems and software that touch on clinical trials, such as those for clinical trial management, electronic data capture (EDC), subject randomization, adverse event reporting, and management of regulatory documents. Although some activities in SOPs can be described generically without consideration of specific systems being used, such as protocol development, other activities such as design of electronic case report forms are significantly impacted by the system being used. The use of EDC versus paper-based data collection had an impact on many SOPs, and companies that still conduct both kinds of trials generally have to have some procedures whose scopes are specific to the two methods of data capture. Even when EDC is used for all studies, different EDC systems can impose technical restrictions or requirements that are essential to a given process and so impact procedural documents— and if this impact is not at the SOP level, then it is certainly at the level of work instruction.

It may even be true that versions of the same system will impact SOPs. For example, a widely used EDC system had required the same set of steps when changing the study database whether the change was to modify fields and forms or to modify the edit checks used to assess correctness of the data. A new version of that system introduced a feature that permitted a faster, less complex process to be used when only edit checks were to be updated. In order to make use of that new feature, SOPs on study database changes had to be updated in order to prevent the faster, simpler edit check update process from being out of compliance with the existing SOP.

When a software system upgrade is planned, companies know they have to satisfy the requirements of *21 CFR Part 11* and good clinical practice; they know they need to fully validate the upgrade to show that it works correctly and consistently and does not impact existing data. Some companies, however, forget to add an assessment of SOPs and supporting documents to their implementation plans for these upgrades. Updating of SOPs to reflect changes in software may be required if staff are to use the systems properly and still remain in compliance with effective procedures.

Procedures drift and improve

The SOP that is put in place when a procedure is new may have been piloted and tested, but to some extent it may still be an educated guess as to what will work for all studies to which it applies. As staff members with responsibilities in a given procedure become more familiar with a process, they may identify improvements, inefficiencies, and gaps. New staff joining the company may have seen other ways to do something and want to introduce those new ideas. At some companies, often ones that are growing quickly, this can lead to an uncontrolled situation where SOPs are (mostly) followed but new and alternate methods are introduced. These changes in procedures can be dangerous to compliance even while they improve business efficiency!

All functional groups should set an expectation of compliance and support ways of assessing, implementing, and embedding new ideas while still being compliant with current SOPs. This expectation may mean filing prospective deviations (see Chapter 12) to allow the process changes to be assessed, but in some extreme cases it may mean retiring some SOPs and starting from scratch.

Organizations reorganize

It is truly surprising how often companies change the names of functional groups in Clinical Deveopment, their organizational structure and job titles. As senior management changes, they bring with them strong feelings about what functional groups should be called or how the organization should be structured. Mergers and acquisitions also result in organizational and job changes for at least one of the companies involved, and sometimes both. The two cases in Example 1 show how an organizational change together with an inadvisable use of a department as a responsible party (see Chapter 9) impacted existing SOPs.

Example 1

At one large company, Clinical Data Management reported into Clinical Operations, but after a merger with another company the Clinical Development senior leadership decided to have it report into the Biometrics organization. This change raised issues because existing SOPs for the original company referred broadly to certain activities applying to Clinical Operations—did those now apply to Clinical Data Management or not? Another company had a similar problem where Clinical Research (Clinical Science) was originally part of Clinical Operations but was then split out into its own department. There, too, staff asked whether activities in SOPs for "Clinical Operations" now applied to the Clinical Scientists or not.

In both of these cases, the only good solution was to update all the SOPs. In the first case, updating the SOPs was more challenging because they were legacy SOPs and were being replaced by new integrated SOPs for new studies. The problem was that the old SOPs typically still applied only to ongoing studies started under the old organization and so were lowest on the list of priorities for the Controlled Document group to update.

Unfortunately, there is no ideal solution to this problem of how to manage SOPs when an organization changes. Some companies *require* updates to SOPs when group or job names change to ensure responsibilities are clear—and this is clearly the right solution, but it may not be possible to carry it out in a reasonable timeframe. Some companies permit the SOP to remain as is, if responsibilities did not change (as in Example 2), and other companies allow minor updates, which do not trigger retraining, to be made to the SOPs in this case. At a company where two large firms were merging, SOPs had tables as attachments listing all the roles in the document, what that role was generically, and what company job titles might take that role. They later moved to including the role names in a company-wide glossary. Because roles and responsibilities in SOPs are a commitment of resources and are closely tied to approval of the process and training, the accuracy of the organizational references matter. Staff that work on maintaining SOPs must keep this in mind when reorganization is planned or a job-title realignment is announced.

Example 2

At a company undergoing rapid growth, the name of the role of the data manager kept changing: "clinical data manager" to "project data manager" to "study data manager." The job description did not change, the responsibilities did not change, but the job title did. In this case, the SOPs in effect had a variety of names for this role— clearly confusing to new staff but because the group performing the work did not change, the Controlled Document group decided that the SOPs did not need to be updated until the process itself changed.

Documents change

In addition to the changes in the environment, other controlled documents and department-managed documents will change and impact SOPs.

Updating templates and forms

In our introduction to controlled forms and templates in Chapter 2, we learned that forms and templates have to be associated with an SOP or a supporting document and also must be specifically mentioned in that SOP or supporting document. However, when a form or associated template needs to be changed, it should be possible to do so without updating the SOP and most, though not all, controlled document systems support this idea. Permitting independent updates to forms and templates may introduce just one problem: because training is based on the SOP and there is rarely separate training on the associated forms and templates, the question does arise of how to make people aware of the change to those associated documents. In theory, everyone will have been trained to always get the latest copy of the form or template from the controlled document system, but for frequently used forms in the real world people will typically download a copy and use it repeatedly rather than downloading it each time. A good communication plan needs to be in place for changes limited to forms or templates to ensure that everyone who needs to be aware of the change knows that it has taken place.

Managing cross-references

Procedures in Clinical Development are not silos of activity that have no connection to other processes being carried out in the organization and so the SOPs governing those procedures rarely stand completely alone. In order to make the entire process clear for users and inspectors, SOPs often refer to other SOPs. Those references have to be kept current to avoid issues of compliance, which means that sometimes a change in a single SOP that appears to be manageable cascades into updates of several SOPs—with all the additional review and work that entails. This cascade of impacted SOPs sometimes arises when a referenced SOP's scope is significantly changed and frequently arises when a referenced SOP is retired. The more specific the reference, the more likely it is that the referring SOP will require an update as demonstrated in Example 3.

Example 3

Consider the situation where an SOP on coding, let us call it CC-001-SOP, is modified. In the past it had used a single form, CC-001-FRM-1, for approval of both adverse event and medication coding, with check boxes indicating what the approval was for. Now, the Coding group finds that it needs to split out approval into two separate forms. CC-001-FRM-1 is retired (the general practice to avoid confusion) and is replaced by CC-001-FRM-2 for approval of adverse event codes and CC-001-FRM-3 for approval of medication coding. If the study closeout SOP, DM-010-SOP, says that all adverse event and medication coding has to be reviewed and approved according to CC-001-SOP, there is no problem and DM-010-SOP will not need to be updated. If, however, DM-010-SOP says coding has be reviewed according to CC-001-SOP using form CC-001-FRM-1, then DM-010-SOP will have to be updated.

The biggest compliance problems arise when an SOP is retired or when a form or template referenced directly is retired. If an SOP no longer exists, all the SOPs that refer to the retired controlled document need to be updated. This situation occurs all the time and is complicated by the concept of business process owners (BPOs)—which so far had only been an advantage. In Example 3, the Clinical Coding group is the process owner of the coding SOP and Data Management is the BPO of the closeout SOP. Clinical Coding can initiate the update of the coding SOP, but if the forms are mentioned in the closeout SOP then Data Management would have to initiate the update of the study closeout SOP. Unless the Controlled Document group requires that the new SOP versions be released as a packet, it is unlikely that the two groups will be able to time their updates to release together unless the change to the data management SOP is very strictly limited to reflect the change in forms. (Example 4 shows what can happen when significant changes impact a web of cross-references.) It may be necessary to create an SOP deviation (see Chapter 12) that all impacted groups train on, in order to ensure that everyone knows what forms to use.

Example 4

Updates of references can create a logjam of SOP updates that can paralyze the system. One group in a mid-sized company had released a suite of SOPs all around the same time. They all came due for review and revision around the same time and the company had significant updates to make to many of the SOPs. There were SOPs to split, SOPs to retire, and lots of detailed cross-references. They were stuck for more than six months with all SOPs in updated drafts but none moving forward to finalization because the order of release would make the *effective* SOPs out of compliance. Finally, the BPO identified the key SOP to update first and worked with the Controlled Document group to document in a prospective deviation the fact that the cross-references would be incorrect for a period of time. Training was provided for each SOP rollout explaining what to do about the cross-references.

The discussion of SOP templates in Chapter 9 recommends that SOPs list the actively referenced SOPs in the *Reference* section—with or without hyperlinks to the actual documents. That is just a start; to maintain SOPs, the cross-references also have to be maintained in their own list outside of the SOPs. A surprising number of companies do not maintain a cross-reference list for their controlled documents because the controlled document system does not support it. Whatever the cause, companies will find that they have to manually create cross-reference listings themselves using spreadsheets in order to help the BPOs determine what other SOPs have to be reviewed or what other BPOs have to be notified. In fact, as we will see in Chapter 14, Controlled Documents and departments that manage documents generally have to resort to creating a full index of SOPs manually, not just a cross-reference list, to have ready access to the information needed to help them manage their documents.

Initiating an update

Any one of the changes mentioned above, or some other factor, will at some point lead BPOs to decide the time has come to make an update to a controlled document. Companies have different requirements and options for initiating an update. The reason for the update and the extent of the update will impact how quickly the modified document(s) can be posted.

Change requests

Most companies require that a formal change request or change order be filed before any controlled document is updated, but Controlled Document groups use those change requests in different ways. Some groups use a change request as a planning tool to know when their own resources might be needed. It also serves to "reserve" the documents listed in the request for update so that they are not being modified by two different teams. In these cases, the request also triggers the release of an editable copy of the document (something that is not generally available) to the requestor. Other Controlled Document groups use change requests as a way of collecting update recommendations or needs until a critical request comes in that triggers the update process. In this case, the Controlled Document group and the BPOs would review the change requests at regular intervals to assess whether enough requests have come in to warrant an update, even if none of them is of a critical nature. Both of these approaches make good business sense, but in the real world they rarely ever work quite as intended.

It is actually rare for someone outside of the BPO's functional group or department to file a change request. Employees outside of the BPO's organization would not generally notice a change from the categories discussed earlier in this chapter and think to file a change request. Change requests generally come from staff in the BPO's organization who are responsible for managing process documents (those who administer department-managed documents, for example) or from a BPO representative working on an initiative that involves revisions to the procedure in question. Because changes originate from within the BPO and because the BPO often has the most recent editable version already or is creating a completely new version, change requests are rarely filed in advance. If they are, the change description may be very vague because it was filed in advance of the work being done. It is very common for the BPO representative to begin an update or new document and only later, when contacting the Controlled Document group to schedule a review, to be told, "You must immediately file a change request!"

Given that the BPO generally has to be *asked* to file a change request, does it still have value? Even when the Controlled Document group has to ask for a request to be filed, it can still be used as a planning tool and to trigger some of the initial analysis of cross-references that are required (see above). Used this way, regular review of change requests is not required but the Controlled Document group can still contact the BPO point of contact when a request is filed by someone else in the BPO's organization. This is the approach taken in the example SOP of SOPs in Appendix 2.

Major, minor, and administrative changes

Factors that can make BPOs hesitate to make quick SOP updates include the time it takes to go through the rounds of review and approval and the training requirement that is triggered with a new SOP version. As we saw in Chapter 10, whenever an SOP is cross-functional, those functions have to be given a chance to review and comment on any changes, the feedback has to be evaluated, and then approvers representing those functions have to be given time to register their approval. In Chapter 15 we will see that most companies require a minimum training often called "read and understand" or "read and acknowledge," where all impacted roles must read the SOP and provide an electronic signature saying that they have done so. Some companies also include a quiz or knowledge check. When an SOP is highly cross-functional, making too many people read it causes irritation across departments. One way to mitigate this situation and encourage the smaller updates such as might come about for changes to referenced SOP or role names is to permit certain kinds of changes to an SOP version that do not trigger the retraining requirement. In addition, these kinds of changes generally come with reduced review and approval requirements, which also make them faster to post.

There are two possible types of changes that would not trigger retraining: administrative and minor changes. Some companies support both of these, some support only one of the two, and some support neither of these, always requiring full up-versioning. These definitions address requirements in both training and approval:

- *Administrative change*: a change to *information* in a controlled document to make a correction that does not change any procedure or responsibility. Examples include updating hyperlinks and correcting email distributions lists. Grammatical errors may also be corrected. Approval outside of the Controlled Document group is not required for administrative changes and retraining is not triggered.
- *Minor change*: a change to a controlled document to clarify the procedure but not significantly change the activities or responsibility. A minor change is by definition a change that, in the estimation of appropriate subject matter experts, will not require retraining.

An example of a minor change is changing the *name* of a particular group mentioned in the SOP but not the responsibilities associated with that role. Correction to a cross-reference may also be a minor change. A minor change must be approved only by the BPO contact who attests that retraining is not warranted.

- *Major change*: any change to a controlled document that includes substantive modifications to the activities or responsibilities. Retraining is always triggered for major changes and approval by representatives of all impacted functions is required.

Controlled Document groups can easily identify when a requested update is an administrative change, whether or not they are familiar with the procedure, and so are generally comfortable signing for the change themselves. Companies that support administrative changes generally do not increase the version at all, because the content did not change, but may list the update in the version history or in the controlled document system. An administrative change may not even require an email notification.

Identifying minor changes is more challenging because they fall into the "you'll know it when you see it" category as being of a nature that does not significantly affect the way the procedure itself is carried out or the responsibilities so that retraining is not necessary. In this case, the Controlled Document group has to rely somewhat on those most familiar with the procedure to make the call, so approval by the BPO representative is appropriate. Minor updates are usually identified by increasing the version number by a decimal place (e.g., from 1.0 to 1.1). An email notification is generally appropriate. When in doubt or if there is any question about retraining, falling back to the process for a major update is appropriate. Example 5 provides a case where allowing minor changes could head off serious compliance questions.

Example 5

The Drug Safety group had an SOP that included procedures on serious adverse event (SAE) reconciliation. This SOP was referenced in several other SOPs that were owned by groups other than Drug Safety. That original SOP was retired and the details of SAE reconciliation were moved into a work instruction supporting a very general SOP. The SOPs that had referenced the original SOP were not updated; leaving new staff to wonder where to find this critical information—presumably inspectors would have the same question. If the company's Controlled Document group had supported minor changes, references to the original SOP could have been replaced with the new work instruction name without a full review cycle for all of those additional SOPs that now could be called into question.

Keep cycle times short

The biggest barrier to updating an SOP is the time needed from draft to posting for training. If we include a subject matter expert review, functional review, cross-functional review, and compliance review each one week long and then add in update times and approval times, we start to talk about nine and a half weeks from final draft to posting for training—but only if everyone meets their deadlines (see Chapter 10). If the change is simple or there is no cross-functional review required, the time can maybe be cut down to two months. Two months at a minimum can seem like a long time for what might appear to be a minor update, so functional groups or BPOs will tend to wait "until we have a bigger change." They should be strongly encouraged to proceed with the change anyway. It is always possible that during review participants will notice additional updates that need to be made and this should be considered an advantage of the review, not a disadvantage!

As we have seen, the author of the change to the SOP typically carries out the functional reviews, whereas the cross-functional reviews are sometimes managed by the author and sometimes by the Controlled Document group. The compliance review is always in the hands of the Controlled Document group as is the approval and posting process. The Controlled Document group therefore has a significant impact on the scheduling of an SOP release. At one company, although the Controlled Document group indicated that a week was needed for an initial compliance review, in fact they were understaffed and it generally took three to four weeks. That extra time and the unreliable estimates that their SOP authors received on how long the end of the process would really take had a significant negative impact on groups making a push to update key SOPs whose rollouts involved several complex, interlinked steps.

Regular SOP review

Once an SOP is thoroughly embedded and in use, the pressures from environmental changes listed above will start to come into play. The longer the SOP is out, the more little changes will add up. Although keeping the cycle time to do an SOP update down is one way to encourage keeping the SOP current (and users in compliance), Controlled Document groups should also implement a required review time for SOPs; it is industry standard to do so. Most companies require that all SOPs be reviewed regularly and that periodic review milestone is usually two to three years after an SOP's effective date. Unfortunately, even when a company's SOP of SOP requires regular review, these same companies do not do a good job recording the reviews and many do not force action if the SOP is not reviewed at the required time. A radical change to this practice may greatly benefit compliance.

Current practice

It is very common for the Controlled Document group to tell an SOP's BPO contact person that SOP review is due and that the BPO should assign someone to review it. Cross-functional review is rarely enforced at periodic review and, in fact, thorough functional review within the BPO's own group by people who actually do the work is not common. Quite often, someone in management reads the SOP and says, "Yes, it is fine" or "It is close enough." The Controlled Document group records somewhere that the review has taken place, and typically the version number is not increased, because no changes were made, nor is the effective date changed. If users of the SOP cannot access the review information, they will not even be able to tell the last time it was assessed. Example 6 is an actual example of what can happen with periodic review that is not really a review.

Example 6

At one large company a detailed and lengthy SOP covering a wide range of data management activities from startup to closeout was kept in effect for seven years. This was during a period when some studies were still using paper case report forms but more and more studies were being conducted using EDC. The SOP was supposed to govern both kinds of studies. It is highly unlikely that this long and detailed SOP did not have any changes during that entire time! In fact, this SOP was never updated; it was replaced (piecemeal) by new SOPs written specifically for EDC.

Recommended practice

With all the environmental pressures, it is extremely unlikely that there will be no changes required after two years. Even if an SOP was written at a very high level, with few details and no system dependencies, after two or three years the regulatory environment will have evolved and the organization may also have evolved. Even if no significant process changes are necessary, there surely could be clarification in wording or additional attachments that could help explain the process more clearly. By taking this all into account, we can come to a radical recommendation: Controlled Document groups should *enforce an SOP update after two or three years; not just call for a review*. Even if the changes made are small ones that do not have a substantive impact on the procedure, the revision will force the SOP into full and proper rounds of review, increase the version number, and update the effective date.

If, for some reason the BPO has to, or wants to, delay the review and revision, the functional group should be required to file a prospective deviation to the SOP of SOPs (or whichever SOP governs review) and provide a time limit by which they will complete the review and update. This recommendation brings up a very practical and important detail about timing. As we have seen in previous chapters, the full cycle time required for revision of an SOP is quite long and is very dependent on the complexity of the change, level of cross-functional involvement, and availability of compliance reviewers. Because this cycle is so variable it must be factored into the requirement for periodic updates. If the update has to be completed *by* the two-year anniversary of the effective date, it would have to be triggered well in advance. It would be more practical to trigger the start of the update two or three months prior to the anniversary (and this would be an administrative task for the Controlled Document group) and put a time limit on it of posting for training one month after the anniversary. This timeframe will allow the work to get started far enough ahead that it could conceivably be completed by the anniversary and the complexity of the change should be clear by that time. If it is a very complex change or involves a package of related SOPs and other documents, the BPO would know by the anniversary date if an extension will be necessary—plenty of time to file a deviation by the one-month post anniversary date.

SOP of SOPs

The SOP of SOPs cannot provide the kind of details found in this chapter regarding all the different reasons that SOPs need to be actively maintained; for large companies it may be worthwhile to create a manual with examples and advice for common occurrences. The SOP of SOPs *can* explicitly require that the Controlled Document group maintain a list of cross-references and other key information for controlled documents. It can also require that change control forms be used and detail the circumstances under which they must be submitted.

Keeping cycle times short has an impact on keeping SOPs maintained and maintaining compliance to them but it cannot realistically be codified in an SOP. This fact has to be acknowledged by the philosophy and approach of the Controlled Document group and provided to functional group authors of SOPs to be taken in consideration when planning SOP updates. The SOP of SOPs can address the question of whether or not administrative and minor updates will be supported (and what kinds of changes will be categorized as minor) and what the implications are for review, numbering, and approval. The example SOP of SOPs in Appendix 2 explicitly permits all three kinds of changes.

The most important topic in this section, periodic maintenance review of SOPs, must be included either in the SOP of SOPs or in an SOP of its own. There is so much value in enforcing an actual revision, rather than just a review, no later than every two years for a growing company or three for a larger company, that all Controlled Document groups should consider this approach. When implementing this requirement, take into account the lead time required for a revision and build an appropriate window around the anniversary and require a deviation if the revision window cannot be met.

section four

Helping staff follow SOPs

chapter fourteen

Finding SOPs

People cannot follow SOPs if they only read them once when they started their position or when the SOP was released. People doing the work will need to go back and refer to the SOP in the future. In cases where the supporting documents are the tools people need to perform their work, they will need to find the right supporting document, and this is especially true of forms and templates. At companies where SOPs for Clinical Development are just given sequential numbers (SOP-023659, SOP-023660, etc.) and appear as a long list in the controlled document system, finding applicable SOPs can be challenging and even annoying. Using "smart" identifiers that do some level of grouping to the business process owner (BPO) can help, but functional groups typically find themselves having to create different ways of accessing all the SOPs that their staff needs. This chapter will look at different methods for making it easier for all users to find SOPs when they need them.

Controlled document identifiers

Controlled document identifiers provide a handy shortcut to finding and referring to a document. They also stay the same, even if the document title is modified to reflect changes in the SOP's scope over time. A system for arriving at document identifiers is a seemingly prosaic topic, but it is fundamental to a well-established and user-friendly SOP system. The worst system is a multidigit sequential number such as *SOP-0162003*. People will rarely be able to memorize the random association of the number and topic and are unlikely to refer to it that way, much less find it again. They will be forced to ignore the number, refer to the topic of the document, and hope for a robust search engine in the controlled document system. Better to provide users with a way to identify those SOPs that are most likely to apply to them and to keep the sequential numbers to a minimum. We can do this by including the BPO in the identifier, so that SOPs whose BPO is Drug Safety would have DS, Clinical Data Management would have CDM, and so forth. Because people are most likely to need SOPs from their own group and the groups they work closely with, they can narrow down the selection easily this way. If we add a sequential number but keep it to two or three digits, people will have an easier time remembering the numeric portion. Now we just have to put them together.

Companies that do include the BPO in the SOP ID generally use the order *DocumentType-BPO-Number*. Consider the atypical approach of putting the segments into the order *BPO-Number-DocumentType* so that a Drug Safety SOP would be DS-001-SOP and a Clinical Data Management SOP would be CDM-001-SOP. Using this unexpected order has a very powerful outcome. We can now assign the same number that the SOP receives to all of the supporting controlled documents associated with it—remembering that work instructions have to be associated with or support an SOP and forms and templates have to support an SOP or work instruction. When these documents identifiers are sorted or searched for, the search will return all of the closely related documents (see Example 1).

Example 1

Consider the case where a Drug Safety (DS) SOP happens to cover both coding review and serious adverse event reconciliation. Because these are two important topics, there are two detailed work instructions—one for coding review and one for reconciliation. Each of those work instructions has a form that is used to obtain approval prior to finalizing the data for an interim analysis or study database lock. We can number the five related documents this way:

- DS-101-SOP
- DS-101-WI-1
- DS-101-WI-2
- DS-101-FRM-1
- DS-101-FRM-2

The work instructions just get an additional number (or it could be a letter) added on, and even though the forms are assigned to different work instructions they are also assigned additional numbers sequentially. This is simply to avoid over-optimizing and making the numbering scheme too complicated, and we still retain the desired effect of having the related documents group together. (In the worst case, the user would have to open both forms to find the right one to use to get approval for SAE* reconciliation.) Now, if we know the SOP number, then we can easily get all documents associated with this SOP by searching for documents with IDs that start with "DS-101."

If manuals are supported by the controlled document system, the company may or may not have a requirement that they be associated with an SOP. If they must be associated with an SOP, they would follow the same numbering system with the identifier MAN. If they do not have

---

* Severe adverse event.

to be associated with an SOP, they can have a simple numbering system where each MAN is assigned a sequential number in the BPO grouping—perhaps the 900 series in our example. So if Drug Safety happened to have a manual on SAE reconciliation that had screen shots explaining how to generate the reconciliation reports, it would not be numbered DS-101-MAN, it would be numbered DS-900-MAN and the next manual on any topic owned by Drug Safety would be given the identifier DS-901-MAN.

We now have an approach that is an aid to users and is especially useful when the most typical way of accessing SOPs is from a list returned by a search or filter in the controlled document system. Individual functions or departments often find other ways to help their staff find the documents they need. One very common approach is to have an internal website that lists the relevant documents categorized in ways that make sense to the group. Another less common approach is to create a detailed index.

Internal web pages

Because most employees prefer a more customized approach over searches in the controlled document system, departments or functions frequently create web pages on their internal web site and provide links to the documents their staff needs, organizing those links in ways that make sense to business process. Clinical Operations and Clinical Data Management tend to create groupings by organizing documents according to those needed for study startup, study conduct, and study closeout. Biostatistics may group documents around their key deliverables such as the statistical analysis plan, randomization, unblinding, independent data monitoring boards, and clinical study reports as described in Example 2. These pages have links to SOPs, work instructions, forms, and templates in the controlled document system and *also* all the department-managed documents in one place. Some companies might even include links to training materials in the same groupings. The documents listed will also go beyond those owned by the department. Using the example above, Clinical Data Management might have links on their web page to the documents related to SAE reconciliation and coding, for which Drug Safety is the BPO but where CDM plays an important role.

Example 2

Biometrics department managers at one company repeatedly got complaints from staff who reported that needed documents were hard to find. The process group in the Biometrics department was tasked with coming up with a better way for the staff to access documents and decided to create a web page. Restrictions on the web pages available to them limited their options to very simple groupings or lists.

In researching staff needs, they discovered that new staff members preferred to have documents grouped by study startup, study conduct, and study closeout because by reviewing the documents found in a grouping, they were able to get an overview of all the tasks that needed to be accomplished during that stage of a trial. On the other hand, more experienced staff preferred to have the documents grouped by topic, such "coding" or "randomization." The working group could not make a decision as to which was the best approach until the system programmer for the group recommended that there be a button on the webpage that the user could click to change the display groupings from one to the other. Everyone agreed that this was the best approach to the problem. The implementation of this approach was made easier because the Biometrics process group *already* had a detailed document index that contained all the metadata for the documents of interest, including both study stage and main topic, so no additional preparation work was necessary in order to be able to present the information both ways.

These department web pages have links to the place the documents are stored, not copies of the documents. This is especially important for items from the controlled document system. It is a dangerous practice to keep copies of controlled documents outside the system for more than quick reference, because updates could happen and the old versions could be confused with new versions. When documents are downloaded or printed, some controlled document systems will place a watermark (see-through print) across the pages saying that it is an uncontrolled copy that is being viewed or printed.

Indexes

Searching in the controlled document system and going to department web pages are the methods most frequently used by staff looking for a specific document to support study work. There are other reasons to search through or find SOPs or other documents that involve more of the attributes, or metadata, of a document, such as what other documents reference it, that are more easily handled through an index table. The Controlled Document group may maintain such an index for controlled documents (through the controlled document system or manually), but departments will also need to maintain lists of their own department-managed documents. Even though these document indexes have to be maintained manually, the effort pays off many times over.

Useful index columns include the following:

- Type of document (SOP, work instruction, manual, template, form, example, checklist, etc.)
- Whether it is controlled or department-managed

- Document identifier (if any)
- Title
- Version
- Last effective date
- List of other documents referenced (both controlled and department-managed)
- Business process owner (if it is not included in the identifier)
- Study period (startup, conduct, closeout)—perhaps allowing for more than one
- Main topic (see discussion above)—perhaps allowing more than one
- Roles impacted or having responsibilities

This kind of information makes it possible to quickly respond to the need to identify which documents reference a specific document that will be updated, who needs to train on a given document, and even reporting to management on how many documents were updated in Q1. As seen in Example 2, the index can also be used to form the basis for providing more user-friendly ways of getting to documents.

SOP of SOPs

The SOP of SOPs may just say that a unique identifier has to be provided and leave it open, or it can provide the numbering scheme, possibly as an appendix. The example SOP of SOPs in Appendix 2 does mention the requirement for maintaining a document index in the SOP of SOPs because of its value, but many companies will consider this requirement to be more of an efficiency rather than a compliance or business need, and so it is best left to a best-practices document or manual.

chapter fifteen

Training on SOPs

SOPs are standard procedures that need to be followed whenever anyone performs tasks governed by the SOP. The only way to ensure that an SOP is followed is to train everyone involved in the task in such a way that they know about the SOP and how it guides the process. Poor training will lead to poor compliance. This chapter discusses several ways to train on SOPs and how to deliver that training to everyone who needs it.

Options for training

The best way to learn about an SOP is in the context of learning the job and the computer systems that will be used to perform the work. But this is rarely done; the most common method of "training" on SOPs is to have new employees read them and sign off (often referred to as "read-and-acknowledge" or "read-and-understand"). Even when read-and-acknowledge is used as the main method of ensuring awareness of SOPs, there are options and approaches that can improve the effectiveness of that minimal level of training.

Read-and-acknowledge

Being able to easily show training records that individually list SOPs to an inspector is one of the main reasons for the overwhelming use of read-and-acknowledge training. If a company had a policy of integrating SOP learning with other kinds of training, they would have to explain this system to an inspector and might be called upon to produce training materials for each course to demonstrate that SOP content was adequately covered. Then, because employees would only get new-hire training from these courses once, the question of how to record training on SOP revisions would also have to be addressed. Training on revisions would probably have to fall back on read-and-acknowledge unless the *delta* training described below was tracked. These difficulties result in just about every company using read-and-acknowledge training as the main delivery system for SOPs.

Even if read-and-acknowledge has to be used for tracking purposes, companies should still consider integrating the relevant SOPs into other required training courses, either instructor-led or computer-based. Understanding the context of the SOP enhances compliance considerably.

We never talk about the *why* in the text of an SOP; training is the place to do that, and when people understand why a procedure is designed a particular way, they are much more likely to follow it.

Quizzes as a form of training

Requiring the reader to pass a short knowledge check or quiz takes the SOP read-and-acknowledge training one step further. Not all controlled document systems (or learning systems that manage controlled document training) support quizzes, and companies that could have quizzes do not necessarily require their use for all documents. If quizzes are used to reinforce learning rather than to see if the reader picked up details, even multiple-choice questions become an additional tool to improve under-standing and compliance.

For SOPs of reasonable length covering one main activity or topic, a multiple-choice quiz would usually have between five and ten questions and require 80% correct answers to pass. Most systems that support quiz-zes support multiple choice with one correct answer; some systems support multiple answers (e.g., answers A and D might both be required for the answer to be judged correct). If multiple answers are not supported, the text of the possible answers provides the options. For example, if the question asks which two roles approve a particular document, then a system that allows multiple responses would read, "Select two:"

1. Data manager
2. EDC* programmer
3. Statistical programmer

In a system that supports only a single response, the question would read: "Select the correct answer:"

1. Data manager and EDC programmer
2. EDC programmer and statistical programmer
3. Data manager and statistical programmer

Quiz software rarely supports free text answers, but advanced learn-ing systems may support more innovative response techniques such as drag-and-drop or the ability to connect two elements with lines.

The best quizzes come from the authors of the controlled document rather than an independent learning group, but those authors may not know how to go about creating a quiz and may start by trying to "trick" the reader or focus on a small detail. Instead, they should start by asking

---

* Electronic data capture.

themselves: what are the three to five most important things I want people to remember when following this procedure? Those three to five concepts will form the backbone of the quiz, with each question tailored to highlight one aspect of those concepts. Ideally, the questions should not require the reader to go back to the SOP, though it must be noted that most employees *do* go back and reread sections of the SOP if they don't know the answer in a quiz, and rereading is not necessarily a bad thing!

Guidelines for successful quiz writing include the following:

- Questions should be simple rewording of the text in the document.
- Do not expect the reader to remember lists of responsible roles, documents, form numbers, or similar details. Try instead to pick one or two key items from important lists to highlight.
- Keep true/false questions to a minimum; no more than one true/false question per every five questions in the quiz.
- For multiple-choice questions, provide more than two choices to select from.
- Try to make most of the possible answers for multiple-choice questions plausible.
- If the system supports more than one answer to satisfy a multiple-choice question, specify how many selections are required. For example, "Select the three forms that must be collected from the investigator prior to the study initiation visit" rather than "Select all the forms that must be collected …"

Example 1 illustrates how some of these guidelines function in use.

Example 1

An SOP explaining the procedure for processing those select subject case report forms (CRFs) required for inclusion in a regulatory submission had an important prerequisite: the PDF versions of the completed subject CRFs were expected to already be available in the trial master file (TMF). (The task of filing completed subject CRFs for each study in the TMF was covered in another SOP on study closeout). Because this is an important prerequisite, we can reinforce it by including a question in the quiz that reads "What must be available in the TMF before CRF processing can begin?" The answers might be the following:

1. Clinical study report
2. Protocol
3. Subject case report forms
4. Sample case report forms

Notice that all of the choices are plausible items because they all go into the TMF. Choice A, the clinical study report, does indeed need to be available for the submission but it is not a prerequisite, and at this company CRF processing can begin before the final clinical study report for the final study to be included has been completed. Choices C and D both use "case report form," but only the *subject case report* forms, choice C, must be available before processing of those forms for inclusion in the submission can begin.

If quizzes are used, thought has to be given to what to do if the reader fails the quiz. The most typical approach is to do nothing other than have the reader take it again (presumably after rereading the document, but this is not enforced). There may be a set number of repeats permitted, but what then? Perhaps the reader is required to meet with someone in person to discuss the SOP. Most of the time, readers will pass the quiz since, really, that is the idea. It can be very useful to the Controlled Document group and/or the authors of the document and quiz to know some metrics about the quizzes. How many tries do people need in order to pass? What is the distribution of correct answers? What questions are most frequently missed? The value of metrics and feedback of this kind is in making improvements: if the information is available, the author should use it to make improvements to the document (or quiz!) at the next revision of the SOP.

Just-in-time training

When read-and-acknowledge training (with or without quizzes) is used, it is still most typical to assign any SOP a newly hired employee could possibly use in their work all on the day they arrive. Often 10, 20, or more SOPs will be assigned that have to be completed within two weeks or a month of joining the company. Employees who arrive with some experience in the work will bring with them a conceptual structure into which they can fit the SOPs they are reading. This technique does not work for people new to the position, who read the SOPs with little understanding of the context. It can also be a problem for people who have not used a company's software systems if those systems impact the procedures in the SOP; they will not have the *why* that they need to put the SOP tasks into context.

An improvement on this system of assigning everything all at once is just-in-time-training, in which the employees and their managers are responsible for ensuring specific training of any kind is performed before doing work in the area to which the training applies. Just-in-time training has become quite common for many kinds of training and usually involves instructor-led or computer-based courses. The employees have the lists or curricula for major tasks and the managers provide oversight by reviewing reports of what curricula elements have been completed for all of their staff. Few companies, though, risk this for SOPs. One company that did had a very clear demarcation point: employees conducting paper trials did not need to take courses or read SOPs for the new EDC system that was being introduced until they were scheduled to start or take over a study using that system. Employees who joined the company and only worked on EDC studies did not have to train on SOPs associated with the paper studies. This policy was true for all roles in Clinical Development, rather than in just a single department. Most companies will find it hard to provide a core list of essential SOPs and then rely on their staff to read and acknowledge the rest, but even when read-and-acknowledge happens all at once, just-in-time refreshers can be a huge benefit for employees as demonstrated in Example 2.

Example 2

One company created a short training presentation on how to go through the steps for study database lock, taking into account the requirements for their EDC system and the integrated applications that company had added on. This presentation was given to each study team at the study close kick-off meeting and was scheduled by the study project managers who knew it was a requirement. This just-in-time training was very well received by all members of the study teams, especially those working on Phase III studies where they might not encounter study locks for several years at a time. In addition, further training was available to data managers on how to use custom reports to check for required conditions such as the presence of all principle investigator signatures and how to most efficiently set the lock on records in the system. The data manager was responsible for signing up for this training.

Who has to train on which SOPs?

No matter whether SOPs are all assigned when an employee starts or whether they are self-selected and overseen by management, there has to be a way to determine to whom any given SOP should be assigned. At the most basic level, as we have seen in previous chapters, training assignments are based on who has responsibilities in a given process. Very small companies can usually just assign everything to everybody, since the groups in Clinical Development will work very closely with each other and may take on each other's task. Small but growing companies can assign by functional group so that any activity that impacts Clinical Operations is assigned to everyone in Clinical Operations, regardless of title or job description. But there are nuances in responsibility and assignments that quickly come into play in mid- and large sized companies that require more targeted assignments—at this point, they need to develop training curricula.*

Determining training assignments

Companies will have an excellent starting point for determining training if they use an SOP template where there is a *Responsible* section with a RACI or summary of impacted roles or if the SOP template uses the table format with a responsible role for each task (see Chapter 9). By default, anyone listed as responsible for a task will have to train on the SOP. However, if the only responsibility a certain role has is to review and approve a document and that review and approval is initiated or guided by a different participant in the process, it may not be necessary for people in that role to train on the SOP. It is generally a good idea for those who only review or approve to understand how that particular document fits into the overall study conduct even if their involvement is limited, but there are understandable exceptions. It works the other way also: certain groups may request to train on a document even if they are not directly involved, so that their members understand more of the process. For example, the person taking on the role of study leader may want to understand the requirements and handoffs involved in severe adverse event reconciliation even if the study leader is not actually involved in the activity. Keeping track of the required and requested training requires a curriculum for each role.

Developing training curricula

At a small or mid-sized company, the curricula would be managed by creating a spreadsheet with all the Controlled Documents applying to Clinical Development down the rows, and all the functional groups (Clinical Operations, Clinical Science, Drug Safety, Biostatistics, Statistical Programming, Clinical Data Management, etc.) as column headers. An "X" in a cell means that function has to train on that controlled document. The rows can come directly from the controlled document index recommended in Chapter 14, when it is filtered by type of document. At larger companies, the functions themselves will want to subdivide into smaller groups so that Clinical Operations clinical research associates receive different training than Clinical Operations study leaders and clinical data managers receive different training than the EDC programmers that also report into Clinical Data Management. Rows in addition to controlled documents, representing

---

* A *curriculum* (plural *curricula*) is a term in education used to describe the list of subjects in a course of study. In Clinical Development, a curriculum would be the list of SOPs to learn and other courses to complete.

required courses, will be added also. The spreadsheet gets more columns and rows as the company grows, but the concept is the same. At global companies with large numbers of employees in each group, the curricula can be broken down even further as demonstrated in Example 3.

Assigning curricula to staff

Once the curricula have been determined, how do they get assigned to actual employees? There are two approaches: everyone with a given job title is assigned automatically to a specified curriculum role or the curriculum role is created or named employees are added to that role as they are hired or transferred to a department. Using job titles works reasonably well if more than one title can be assigned to the same curriculum (e.g., clinical scientist and clinical science associate). The approach of adding individual names to curriculum roles provides a great deal of control over assignment of courses and SOPs as shown in Example 3, but the maintenance effort is higher and probably only makes sense for long lists of required training and a large number of employees in the department(s). This approach is also less appealing if training for SOPs is managed by the controlled document system and other kinds of training are managed in a different learning system, because assigning names to roles in two locations (and keeping them synchronized) would be a burdensome task.

Example 3

At a company where nearly all new data managers started by helping out with query management and other study conduct tasks, management created a training curriculum of *CDM study conduct* and assigned it to new staff. When a manager determined that a data manager had enough experience to help with a study closeout, they were assigned the *CDM study closeout* curriculum. Because study builds required the most experience, being assigned the *CDM study build* curriculum typically came last. The curricula at this company included both SOPs and all other required training for each of those roles.

Employees were associated with a given curriculum by name. All SOPs in the curriculum were automatically assigned immediately when a person was assigned to a role, but completion of training classes was the responsibility of the employee. The SOP read-and-acknowledgement due date was extended to 60 days for new employees to allow them to take appropriate introductory instructor-led or computer-based classes first and then read the related SOPs after completing those. Experienced data managers were assigned to multiple roles from the start. After a while, this company decided that maintaining both *CDM study conduct* and *CDM study closeout* curricula was not worthwhile, and they combined the two but left the *CDM study build* role as a separate curriculum.

It is worth noting that most companies have some SOPs (such as the SOP of SOPs) and corporate policies (such as those governing financial arrangements with health-care providers) that everyone in Clinical Development will need to take. Those SOPs and policies are usually managed as a separate set; they are applied automatically to all new hires and reassigned when they are revised. For very important topics, they are automatically reassigned every few years to ensure employees remember the key points. Documents that fall into this category do not need to be included in the curricula spreadsheet managed by the departments.

Maintaining curricula

Once the initial curricula are established, the list of controlled documents (and other kinds of training) in those curricula must be kept current and assigned to appropriate roles. For new SOPs, the Controlled Document group can make an initial assessment and recommendation based on the responsibilities in the document. But then, just as each involved department or group has to approve their steps in an SOP, they will also have to approve or modify the training assignments. For revised SOPs, a new assessment must be made because changes to the process may impact which departments are involved. Because the departments have to decide which roles must train on which controlled documents and because departments will add on additional kinds of training for each role, training curricula appear in the document hierarchy introduced in Chapter 2 as department-managed.

So, who in the department gets to make the decision on who is assigned what? If the department has a group overseeing department-managed documents and SOP work, proposed curricula changes for new or revised documents can be routed to that group first. That group may or may not have the responsibility or authority to make the decision. At many companies, questions of training are routed to a review team or management team for resolution. When the curricula are active, with updates and releases of controlled documents as well as updates to other required training arising all the time, it may be necessary to route each curriculum for a full review every half year or year, so that the assignments can be assessed as a whole. The impact of seeing the entire table of assigned documents and courses can lead to adjustments in due dates for courses or may lead management to consider consolidating roles or breaking them into smaller, manageable units. At a minimum, this kind of review serves to make direct line managers more aware of the amount of training new hires will have to complete, which should in turn lead managers to have more realistic expectations of when a new employee can begin direct study work.

Training contract staff

Most companies, both large and small, will have some contract or temporary staff. If those people work on-site or report directly to in-house management, it should be pretty clear that they need to have (nearly) the same kind of training as permanent staff. There are variations in the contract model that make the discussion of training on controlled documents interesting, and there are technical restrictions that can introduce challenges.

The first variation is that seen in Clinical Operations when monitoring at a small company is contracted out to a contract research organization (CRO) for studies that are otherwise conducted in-house. In this case, the site monitors may be following the CRO's SOPs and any contact with the sponsor may be mostly through the CRO's project lead liaison to the sponsor. The CRO would document training and the sponsor would only need to check that training was being performed as required. Whose SOPs were to be used for which activities would be documented in both the vendor's contract and in the study's documentation.

The next variation, seen frequently in Data Management Organizations and often called the *functional service provider* (FSP) *model*, has staff contracted through a CRO working remotely but following the sponsor's SOPs on the sponsor's computer systems. In this case: what should the training be, who is responsible for training, and where is training recorded? In this FSP model, the contracted staff usually work on a subset of the activities performed by full-time staff. In the example of Data Management, one vendor may provide staff to assist in study build activities for EDC and another vendor may provide staff to assist in managing lab normal ranges and lab data queries. To provide the most appropriate training, the sponsor will create curricula specific to these vendors. The study build FSP staff may perform the very same activities as in-house EDC programmers and so the curriculum will look the same. But the FSP staff that manage lab data will need only a small subset of the curriculum a full study data manager would receive.

At some companies, the FSP staff is given network access and can get to the controlled document system and the learning system. This is the easiest approach for everyone all around. Even when the FSP provides certain instructor-led training required by the sponsor to its own employees (in a "train the trainer" model), it can be easily recorded in the sponsor's records. Unfortunately, the security policy and/or technical infrastructure in place at some companies means that remote FSP staff cannot be given network accounts and cannot access the sponsor's learning system; alternatively, they can be given network accounts but still cannot access the learning system or controlled document system.

This lack of access means that all controlled documents have to be made available to the FSP using a manual process—and kept current—and the FSP will have to keep the records. Because some auditors do not accept that the training records be kept remotely, companies can find themselves having to request that the FSP send scans of paper records back so that the sponsor can keep a local copy of those records.

The point about having to keep the FSP current with controlled documents is especially important and should not be overlooked. If the FSP does not have access to the controlled document system, then someone at the sponsor must inform CRO or other service vendor contacts *every time* controlled documents are released, retired, or updated. Add to this any essential department-managed documents that the FSP staff require for their work (though this will likely be easier in cases where FSP staff have network access because department-managed documents are typically not stored in the controlled document systems). The management of documents for the FSP vendors used by a department can be a substantial administrative task. This task is made especially complex when the department uses different vendors that provide staff to perform different activities, in which case, each vendor will have to have its own list of documents. This administration is usually left to the functions rather than supplied by the Controlled Document group, and the functions will find that assigning this task to a specific person is the best way to ensure that it happens. That person would also be responsible for maintaining the FSP's curriculum, supplying documents, and periodically auditing the vendor or CRO training records to ensure that all required training is taking place.

Training periods

Most companies follow the approach of posting controlled documents "for training" for a certain period before those documents become effective and must be followed. Two weeks for training is a common time period but the time allotted for training should be flexible. For urgent updates to SOPs, as sometimes happens when an error was introduced in the initial release, the period can be shortened. Training periods are lengthened when the change is substantial in order to allow impacted departments to provide additional training or presentations to their staff in addition to the required read-and-acknowledge of SOPs.

During the training period, employees can be confused about which version of the SOP to follow. Because of this, some companies do not make the new version of the SOP available except through the training system until it becomes effective. This way, anyone looking for the SOP in the controlled document system will see only the one that is still in effect. It is usually necessary to remind all impacted departments of this on each and every release of an SOP because some-one will always go through the read-and-acknowledge and then have a question about the new document and not be able to find it. Some will also want to compare it to the previous version (see also Chapter 11) to better understand the changes.

Delta training and minor releases

Presentations on the changes (sometimes called "delta training" in a reference to the mathematical *delta* symbol for change in a value) are a tremendous tool in improving compliance to SOPs. For revisions to an existing SOP that introduce efficiencies or better reflect practice, a five or ten minute presentation can highlight what *has not* changed and what *has* changed. Those employees who are not present for the presentation can read the slides and get almost as much out of it. For significant process changes that result in a rewrite of the SOP, the delta training will be more substantial and is provided to current staff to ensure they understand the new process. Both small and large changes will have to be integrated into all existing training courses associated with the procedure. All new hires will get that updated training; it is only for extreme changes (such as past changes from paper to EDC) that we want to have current staff complete new-hire training.

For the most part, all current employees in impacted departments will need to reread and acknowledge again any new SOP versions. Some companies support minor revisions to controlled documents for corrections that do not significantly impact the process or responsibilities (see Chapter 13). When minor revisions are supported, they will not trigger retraining requirements.

SOP of SOPs

Although the basics of training on SOPs may be mentioned in the SOP of SOPs, the full discussion is broad enough to warrant a separate SOP on training for controlled documents or an SOP on training records. Wherever training on SOPs is covered, that document should provide the guidance that any role with responsibilities in an SOP should train on that SOP. If a company plans to allow any exceptions to this rule, such as when a role has review and approval activities only or when groups may require an SOP for information, the curricula act as the means of communicating and documenting these exceptions.

The SOP of SOPs usually does sketch out the training requirements for the other types of controlled documents such as work instructions and manuals (templates and forms are usually not trained on separately). The SOP may also reference the training period but that may be relegated to a supporting document. Training of contract and service provider staff and maintenance of curricula are not usually addressed in an SOP but there may be another controlled document that governs providing vendors access to internal documents.

chapter sixteen

Department-managed documents

Although SOPs are the main subject of this book, department-managed documents nearly always play a role in the practical, real-world use of SOPs and are *required* in order to follow many SOPs. The concept of department-managed documents was first introduced in Chapter 2 as part of the document hierarchy, and it has been mentioned in nearly every other chapter that followed. In this chapter, we will look more closely at how the management of these documents is similar to, and differs from, SOPs. The sections below, which mirror the broad sections of the book, provide successful practices that can make these documents a support to—rather than a risk to—SOP compliance.

Founding principles

When to have a department-managed document

Let us review some of the characteristics of department-managed documents discussed in Chapter 2. Department-managed documents have the following characteristics:

- Provide details, instructions, workflows, templates, examples, best practices, and other information that supports controlled procedures and activities around those procedures
- Apply only to the staff of the department that manages them
- Are maintained by department staff and made available through a shared drive or internal web page

Department-managed documents should be used with caution because storing documents in the controlled document system has definite advantages, particularly the built-in version control and tracking of training requirements. The disadvantage of the controlled document system is that it makes getting documents updated harder. This leads us to the three areas where department-managed documents are most useful:

1. When the process is new to broad use
   Documents describing the details of a new procedure for a department should be department-managed to start. Past the pilot phase,

when the process appears to be correct but wider use is expected to shake out any issues, updates are very likely. To put it into widespread use, the document must be readily available to everyone who needs it. When issues are found, they can be quickly updated and released. The department plans to move the document into the controlled document system in the future.

2. Documents that need to be updated frequently

Some templates or guidelines that change frequently should stay department-managed. In Data Management, the clinical data management plan template is one such document. For Biostatistics, it might be the statistical analysis plan. Both of these must reflect changes in the Biometrics environment on short notice. At companies where default text is provided that supports many therapeutic areas and types of studies, these templates may need frequent updates. Guidelines and instructions can similarly change as the corporation's computer systems and organizational structure change over time; the presence of screenshots and examples increases this likelihood.

3. Informal documents that save time

Documents such as best practices, FAQs, and examples may be very helpful to a department but do not warrant tight control.

In general, the more detailed the document and the more likely it is to require updates, the more likely it is to be a candidate for a department-managed document.

What to say in the document

Department-managed documents must align and support SOPs or controlled work instructions to which they apply. This point cannot be emphasized enough: because department-managed documents can be updated more easily than controlled documents, the tendency is to reflect process updates at that department level first. Sometimes those changes are subtle deviations from effective SOPs and may drift further away over time. Before the department-managed document is released for use, someone must compare it against effective, controlled documents to ensure that there are no contradictions.

Duplication without elaboration should also be avoided. Sometimes a group will make the assumption that there *has* to be a more detailed instruction to go with every new SOP. If the draft of a department-managed document does nothing more than restate the SOP without adding additional useful detail, then the document is not worth having.

In this case, the impacted groups should work according to the SOP, provide appropriate training, and only add a detailed document later when a real need is identified.

The best use of department-managed documents is to provide detail on how to carry out a task. The kinds of detail that work well in these documents include the following:

- Specific email distribution lists, server names, and folder paths
- Information about actual tools being used to carry out a task mentioned in an SOP (such as spreadsheet trackers)
- Instructions on how to perform a task in the computer systems being used
- Vendor-specific content (as in data transfer specifications)

Where to put the output

Instructions provided by department-managed documents may result in the generation of forms, documents from templates, logs, trackers, and other output. As with SOPs and other controlled documents, it is important for each document to indicate clearly what to do with this output if it is not already mentioned in the guiding SOP. Some output may be used as a communication tool and so does not need to be retained after the action has been taken. Some output may be retained electronically in the study's folders until after the study report has been written. It is best practice to mention any critical output that must be retained in the study's trial master file at the SOP level.

Writing, reviewing, approving, and posting

Writing and formatting

Department-managed documents that are procedural documents (like uncontrolled work instructions) benefit from a process map in the same way that SOPs do—the procedure is just a level or two more detailed. But many department-managed documents are not procedural and can be drafted as appropriate to the content. Just as in controlled documents, it is not wise to post a department-managed document that has never been used, with the possible exception of FAQs. Everything else should be piloted or used on an actual study or be derived from a document developed for use in a specific study or studies before being made available to broader use.

It should be clear from looking at or opening a department-managed document that it is, in fact, department-managed and not a controlled document. Otherwise, the header requirements for department-managed documents are very similar to those for SOPs described in Chapter 9, and this includes version and effective date. Department-managed documents need to be version controlled, but even here we can introduce flexibility. For many documents, the same kind of version numbering recommended for controlled documents will work just fine. But there are some documents where the release date or last updated date is more useful than the actual version number. This is particularly true of documents such as instructions or templates associated with standards that are updated many times a year.

Other than the header requirements, there are very few guidelines that have to be applied in the formats or document templates. Documents destined to move to the controlled document system once the content is stable can be formatted from the start in a way that is *similar* to the template of the target document to make the later transition easier. That is, if the document will eventually become a work instruction, then using the controlled document work instruction template from the start—with appropriate changes to title and header to indicate department control—will save time and effort once the process is deemed stable.

For documents that will be permanently department-managed according to the criteria described above, formatting should adjust to best present the content. This statement is obvious for templates and forms but should also be the guiding rule for other types of documents. For example, some procedural documents or instructions will have content well suited to a table format like that used for SOPs which has a responsible role and activity, whereas others really need to provide wordy text background. Manuals and instructions need to be able to support screenshots and examples and should not be held to a rigid outline. Even the numbering schemes should be flexible, because some types of documents benefit from strictly numbered sections and others just need general organization into topics. The important thing for the variety of documents managed by departments is to present the information logically and clearly in a way that is useful to the reader.

Review, approval, and posting

Department-managed documents should go through a review process just as controlled documents do. Because content is limited to the department, there are only two rounds: the subject matter expert (SME) round, for review by SMEs other than those who wrote it, and the function or department review. (See Chapter 10) Best practice for department-managed documents is the same as that for controlled documents: using a set review group whose members represent the different subfunctions, types of studies, and geographic locations of the department.

For department-managed documents that come out of working groups with good representation from the department, it can be tempting to skip the department review and just get the documents out, but review here is just as essential as it is to SOPs—perhaps more so. Staff tend to use supporting documents, controlled or department-managed, more than the SOPs in their day-to-day work because those documents contain the useful detail that helps people do their work. The content and the details need to be both correct and clear. Even when a document applies to a single subfunction of a department or function, if it describes hand-offs or expectations involving other subfunctions, then it needs to be reviewed by representatives from those groups in the same way that SOPs that cross functions have to be reviewed by those functions.

Some companies require management approval for department-managed documents, others do not. If managers are part of the review process, there may not be a need for a formal approval step, because the outcome of the review can be taken to indicate approval once the feedback has been incorporated. But, because management approval shows management *support* for a document, obtaining formal approval helps convey the important message that the material is to be considered the current thinking of the department. At large companies, where there can be differences of opinion between subgroups or locations, this evidence of management support may be essential.

Posting department-managed documents means making them accessible through the department's shared server locations or internal web page. A common mistake made by departments is to post the documents without thoroughly protecting them from inadvertent or intentional changes. Department documents can be protected by creating PDF versions of procedural and instructional documents, but templates, forms, and examples need to be available in their native format. One company found that the file protection had not been tight enough and employees had, on more than one occasion, begun editing before they realized they were editing the posted document directly!

Many department-managed documents are just as important, and perhaps more so, to the day-to-day activities of a department as SOPs are. The posting of a department-managed document, therefore, should receive the same kind of communication plan and training assignments as a controlled document. See Chapters 11 and 15 for further discussion of these topics.

Departments often make the mistake of not keeping the older versions of department-managed documents. Retaining versions is especially important for procedural documents, which might be requested during an audit or inspection. It is *not* true that auditors or inspectors can only request SOPs. All documents and training materials used by an organization to perform critical procedures can be requested; it is just that an auditor would typically start with SOPs and only request other documents if there appeared to be a problem with a procedure or if there were a gap in SOP coverage. Be aware that an auditor or inspector might request documents that applied to a specified, past study that is the focus of the inspection or, if the inspection is targeted at a current process such as site monitoring, the current department-managed documents may be requested.

Figure 16.1 lists the required activities associated with releasing a department-managed document. These requirements could be formatted as a checklist to be used to ensure that none of the critical steps in release of a document are overlooked.

Receive near-final draft:

☐ Check formatting, versioning, and revision history
☐ If document is new, assign a document identifier (☐ N/A)
☐ If document is new, add a placeholder in the document index (☐ N/A)
☐ Review document against related controlled documents
☐ Return document to author for revision (☐ not required)

With author, assess training impact:

☐ Add to curricula spreadsheet or assess existing entry
☐ Schedule creation of, or update to, training system entry
☐ Schedule creation of, or update to, training materials (☐ N/A)
☐ Plan and schedule any instructor-led training (☐ N/A)
☐ Route document for management approval
☐ Draft and review posting announcement with author

On posting day:

☐ Archive current version or ☐ New
☐ Move document to effective-document folder
☐ Add effective date to document index
☐ Update or add link to documents web page
☐ Send posting announcement
☐ Trigger training in training system

Figure 16.1 This list of activities required to release a department-managed document applies after the document has gone through required review. The steps may be carried out by a document group for the department responsible for posting, but the activities could just as well be carried out by the subject matter experts or authors responsible for the document.

Maintaining compliance

Deviations and active maintenance

We talked at length about deviations, prospective and retrospective, from controlled documents in Chapter 12 and noted there that at many companies formal deviations can only be filed for SOPs. If there is a systemic problem with the process as recorded in department-managed documents, then getting that information to the department's document group or the assigned SMEs is essential. For those department-managed documents that are destined for the controlled document system, identifying these issues is the very reason for posting these documents to the department first. When such documents are posted, the announcement should include a contact email for communicating issues and roll-out presentations should encourage staff to ask compliance questions or point out problems. If a deviation is significant, the department can advise the study impacted to note the deviation in study documents using the same techniques available to SOP deviations discussed in Chapter 12.

Note that, although deviations from procedural supporting documents can be important, it generally will not be necessary to record when a person deviates from instructions found in reference documents or when a template is misused. Not following instructions is usually just considered a mistake, as is not using a template appropriately, but staff should be encouraged to provide feedback here, too, to the department's document group or SME so that the instructions or template can be improved and the issue included in training, as appropriate.

It is the rare department that schedules periodic review for department-managed documents in the same way that SOPs are reviewed periodically. Typically, the documents being used most are most likely to be updated and one of the key attributes of many department-managed documents is that frequent updates are expected. However, some companies that did not have a path to move department-managed documents to the controlled document system ended up with badly outdated procedural documents still posted, and still on training curricula, but which applied to seldom-used procedures such as those applying to paper-based studies when most of the studies were being run using electronic data capture. Departments in companies that do not support controlled documents below SOPs, who therefore maintain many documents as department-managed, should implement a periodic review to ensure that these documents remain current.

Helping staff use department-managed documents
Finding the document

Just like SOPs, department-managed documents can benefit from a document identifier, but only when that identifier contains more information than just a sequential number. Departments may *not* want to follow a system similar to that described in Chapter 14 for controlled documents. Within a department, it may not be necessary to identify a business process owner so a document type and short number may be enough, with procedural documents being the focus as shown in Example 1.

Example 1

One particularly useful numbering scheme for department-managed documents, devised by a growing company, had seven main topic areas including study startup, study conduct, and management of vendors. Those were subdivided into document types including those for procedure documents, templates, forms, manuals, and training, to create a matrix with topics as columns and document types as rows. The columns received the number blocks 100 to 700. The rows also received blocks such as 110 to 119 for procedural documents in study startup and 170 to 179 for those in vendor management. A template for study startup, related to a procedure for study startup with the number 110, would receive the number 111 and a template related to a procedure for vendor management, with the number 170, would receive the number 171. This way, documents around a similar task had similar numbers. Although the numbering scheme was particularly handy for the people in the department responsible for managing documents, the staff were more likely to remember a few numbers but would normally find what they needed from the matrix on the department web page (see also Chapter 14).

It is amazing how quickly department-managed documents can get "lost." After just a year or two, you might hear, "I didn't know we had a document for that!" Although training, discussed below, helps keep documents from getting lost, a document index is even more useful and should be considered an essential tool in the management of documents within a department. Each department should quickly be able to answer the questions, "How many documents do we have? How many are templates?" And when an SOP is updated, "Which of our documents have to be updated to align with this SOP change?" Departments must not just delete items that are retired; these should be left in the index with a notation as to the retirement date and the reason why the document was retired. As noted above, past department-managed documents *could* be called for during an inspection.

Training

Department-managed documents do no good if staff do not know they are there. Without a training plan, the typical course of events is for a document to be developed and, when it is ready for use, the author or a manager will present at meetings, send email, and answer questions. Everyone working in that department or function *at that time* will know about the new document and when to apply it. However, a few months later, when new hires begin work, all of that communication is lost and the new staff will not know about the document unless they stumble across it or are pointed to it by someone who was there when it was rolled out. This omission may not be a problem for templates or forms, because they are usually referenced elsewhere, but for procedural documents this can be, and has been at several companies, an audit finding. If a department has important procedural documents that it manages, then those must be added to training.

As noted in Chapter 15, read-and-acknowledge is a minimum standard, but a better kind of training is to integrate documents with instructor-led or computer-based training that provides employees with critical information on how to do their work. Although providing training for SOPs only in courses can be challenging due to compliance considerations, it is much more plausible for department-managed documents. For example, if a manual is released that includes both procedures and instructions on how to carry out coding of adverse events and there is a training that covers the same material, it may not be necessary to assign a separate read-and-acknowledge training for that document. When there are revisions in the future, staff who have already taken the training can be notified about the changes with a short delta training or email notification. Major changes would naturally require retraining for everyone. Some groups that oversee department-managed documents have added a column to their document index that indicates how training or awareness of the document is ensured—by read-and-acknowledge, instructor-led class, or computer-based training.

For documents where read-and-acknowledge is sufficient, it may actually be challenging to track this training. Most controlled document systems have either built-in training tracking for those documents or have integrations to corporate training systems. If a document is department managed, it cannot make use of that automatic connection to training. This means that the group that oversees the documents will have to create items in the learning system catalog for all the department-managed documents that need individual training; this group will have to keep those entries updated as new versions are released. There are companies where this is not possible or is burdensome; in those cases training may have to be tracked on paper forms. And as for SOPs, it is critical to ensure training for any functional service providers who need to be aware of and use those documents.

The document on department-managed documents

Just as the SOP of SOPs guides the creation and maintenance of Clinical Development's standard procedures and other controlled documents, a dedicated document should guide creation and maintenance of department-managed documents. The document on managing department-managed documents can describe requirements for and activities around the topics in this chapter including the following:

- Which documents will be department managed
- The review and approval requirements for these documents
- Release and posting expectations
- Dealing with deviations
- Training policies

section five

Additional topics

chapter seventeen

Where to start

Other chapters so far have made it clear that writing and maintaining SOPs is an ongoing project for every company, but where do you start when you are a new company just getting going with running clinical trials? After you write an SOP of SOPs, then what SOPs have to be written first and what do you do if you have to start trials without a full set of SOPs? This chapter provides some approaches for one common growth path for new companies: starting out as a virtual corporation with most of the study work contracted out and then bringing more and more trial activities in-house over time.

When the company is virtual

Most startup biopharmaceutical companies begin work on clinical trials as *virtual companies*, where in-house staff act as the sponsor contact but nearly all the study work will go to one or more contract research organizations (CROs). When contracting out responsibilities in a clinical trial, the sponsor has to be guided by these two sections in *ICH E6 GCP*:

- 5.1.1 The sponsor is responsible for implementing and maintaining quality assurance and quality control systems with written SOPs to ensure that trials are conducted and data are generated, documented (recorded), and reported in compliance with the protocol, GCP, and the applicable regulatory requirement(s).
- 5.2.1 A sponsor may transfer any or all of the sponsor's trial-related duties and functions to a CRO, but the ultimate responsibility for the quality and integrity of the trial data always resides with the sponsor.

These excerpts together tell us that even when most of the work is contracted out the sponsor must still have a system of SOPs in place, and the SOPs should focus on showing how the sponsor will ensure the quality and integrity of the trial data. The MHRA* has written about

---

* Medicines and Healthcare Products Regulatory Agency, the regulatory agency of the United Kingdom for medicines and medical devices.

this situation[*] and provides the example of a sponsor who outsources monitoring to CRO staff who follow their own SOPs—but the sponsor still has an SOP on how to perform monitoring. The MHRA explains that the sponsor should instead have SOPs in place to describe how the CRO's work will be overseen, which might include comonitoring or review of the monitoring reports.

The MHRA also points out that the sponsor must have SOPs governing the contracting process because this is the point where the scope of work and responsibilities are defined. This is essential in order to satisfy the next two sections of *ICH E6 GCP*:

- 5.2.2 Any trial-related duty and function that is transferred to and assumed by a CRO should be specified in writing.
- 5.2.3 Any trial-related duties and functions not specifically transferred to and assumed by a CRO are retained by the sponsor.

When the sponsor and the CRO have agreed to the responsibilities each will assume, the sponsor must then assess whether SOPs are needed to govern the activities sponsor staff will perform.

After writing the SOP of SOPs and creating an SOP template, the company should focus on the contracting process and define what steps are required before working with a service or lab vendor. The SOPs to work on after that are those governing how a vendor's work will be overseen during the conduct of a trial and how data and data analysis will be evaluated. Avoid posting an untried process for these early SOPs by keeping the initial version to a fairly high level. Ensure that you have a secure area to store those SOPs and a way to document training on them. Soon after starting other SOPs, create an SOP on SOP Deviations and aim to have it in place before the SOPs are used on studies.

Small companies taking on activities

As companies grow and bring in more staff across departments, most begin to bring some trial-related activities in-house. Which activities they start with will depend on where staff and resources are strongest. This is the point at which the list of responsibilities from contracts with vendors again plays a role. The sponsor needs to have SOPs in place for all key activities the sponsor will perform—but it is very helpful to know that they do not *all* have to be in place at the same time; in fact, as we saw in Chapter 10, it is nearly impossible to push a lot of SOPs through in a short period of time.

* MHRA. 2012. *Good Clinical Practice Guide*. Chapter 14, Section 14.1.1.

First identify the list of activities that will take place mostly in-house or will be largely directed by in-house staff, even if vendors supply contract staff to perform some of the work. Evaluate that list and review it against Chapter 3. Create SOPs for the most critical GCP topics first; only add or modify SOPs for business reasons as the business grows. For example, when a sponsor takes over writing the protocol where it was previously outsourced, it is not necessary to write an SOP for authoring protocols immediately. The protocol with appropriate approvals and amendments is its own evidence of compliance—the process for drafting, routing for review, and commitments to turnaround times are all more of a business need.

After having identified the critical list, prioritize by timing the SOPs to the stages of the most critical trials. For example, if an important Phase II trial is nearing completion and, if successful, will be a key component of a future submission, then ensure that the study closeout and data integrity activities are well documented. For a critical Phase III trial, key SOPs for each stage of the trial should be in place before the study gets to that particular stage.

What if the SOP is not ready?

When no SOP is available to guide a process, a company can always show that a compliant and high quality process was used by writing study-specific documentation of that process and filing it in the trial master file. For example, if necessary, a trial monitoring plan can include more process information, should an SOP for monitoring not be ready. Later monitoring plan templates might refer to one or more SOPs rather than include the information in the plan. In Data Management, it is very common to use the data management plan to document processes for which no SOPs exist, until such time as the process is standard and can be codified in an SOP. Example 1 shows another example of documenting at a study level, when a company standard is not available.

Example 1

A growing company had a critical Phase III trial, where it was not possible to double-blind randomization because of the extreme difference in delivery methods for the trial medication as compared to the medications for current best standard of care. Even when a study is open-label a sponsor can help prevent bias by avoiding analysis that includes the treatment groups (see *ICH E9, Statistical Principles for Clinical Trials*). Because no SOP for randomized, open-label studies had been written, senior management had the study team write a document that described how the study team would avoid knowledge of the treatment group when assessing serious adverse events, cleaning and reviewing data, and performing initial analyses. As the company grew and further randomized open-label trials were planned, a working group convened to create an SOP to document the best process. The working group used as a starting point the study-level document that had worked well for original trial.

Which SOPs are used by whom?

In the sections above, the emphasis is on identifying those SOPs that are needed for use by the sponsor. When a study or part of a study is fully outsourced to a CRO, the CRO staff will be following their own SOPs. If there is any chance of a question of which SOP was or should be used, the contract or scope of work should explicitly state this. This question of which SOP to follow comes up more frequently when a CRO supplies staff for very specific positions or roles—site monitors are perhaps the most common example. Smaller sponsors may ask the CRO's monitors to use the CRO's SOPs; then, as the sponsor grows and adds its own monitors to a CRO's, they may ask everyone to use the sponsor's SOPs. Whenever the sponsor requests that CRO staff follow the sponsor's SOP, then those CRO staff members must be trained on and have access to those SOPs (see also the discussion in Chapter 15).

Do not post untried procedures

The rule from Chapter 7 that SOPs should not be theoretical still applies, but it has an especially hard impact on virtual and small companies. These companies are growing fast and working fast to meet tight timelines that may mean the life or death of the company. They generally cannot take the time to pilot procedures to ensure they are correct—but they should still not post SOPs that cannot be followed. Two things can be done to address this apparent catch-22: 1) keep the SOPs to a very high level, leaving the details to study-specific documents and plans, and 2) use the first one or two studies as pilots. In Chapter 7, we discussed how pilots typically are run on real clinical trials and so must be conducted with great care and appropriate documentation. This same approach can be used for the first study or two run by a sponsor until the process, specifically as it is implemented at that particular sponsor with its specific computer systems and applications, is more stable.

As the company continues to grow

If the first few trials appear successful, the sponsor will likely move more tasks in-house. The drafting of SOPs continues by identifying what procedures are needed for those activities conducted by the sponsor staff. Sometime around the first submission—perhaps before, perhaps a bit after—if the company is successful, management will hire more staff in more functions. It is at this point that the need for supporting documents (and especially department-managed documents) generally becomes quite strong. To avoid a hodgepodge of documents across all functions, some of which are well managed and some of which are not managed at all, the Controlled Document group should provide information and guidance for the Clinical Development departments on the minimum requirements for critical department procedural documents. As time goes on and there are multiple drugs in the pipeline and several hundred staff in-house, some of the larger groups in Clinical Development, such as Clinical Operations and Data Management, may hire someone specifically to manage SOP creation and revision, oversee the department-managed documents, and act as a point of contact for the Controlled Document group.

chapter eighteen

SOPs during mergers and acquisitions

When companies merge or one company acquires another, there is a period at first when staff from the two companies continue to work under their own SOPs. The topic of which SOPs to follow will begin to come up as an issue when staff from the two original companies begin to work together on a study; then it becomes critical to know which procedures to follow. This chapter explains how the problem evolves and offers some approaches, guided by the principal that either SOP is fine as long as everyone knows what they are following.

At first

Although a single quality management system is desirable, companies can work under more than one for a time, and this situation is always present during mergers and acquisitions. Right before the legal finalization of the event, the two companies will be conducting trials according to their own SOPs and controlled documents. The next day, the companies do not immediately change the procedures they are following; they continue to work under their respective systems. Some smaller companies can miss this important point as seen in Example 1.

Example 1

One company that had acquired several smaller companies over a space of just a few years tried to write SOPs so generally that they could be applied to ongoing studies from any acquired companies. When revising their own data management SOPs, they did not want to create any SOPs that reflected the requirements of the electronic data capture (EDC) system they were using, saying: "What if we acquire another company and it uses a different EDC system that does not require the steps in that order; what will they follow?" It took repeated explanations to various staff members to convey the idea that the acquired company staff could follow its own SOPs for studies that were running at the time of acquisition. Many studies could even continue under those SOPs until they locked. If any ongoing study was expected to be long running, special consideration would be made to identify which SOPs would be followed. It would *not* be necessary to write SOPs that would cover all possible acquired organizations.

For acquisition of small companies, there is typically an effort to get the purchased company integrated with the buying company as quickly as possible—perhaps within a few months. In other cases, such as those where large companies are involved, the merging and integration can take years. Soon after the finalization date, both sides must begin to figure out at which point it will become necessary to document what is being followed for each trial and plan for how this will be done.

The first challenge arises when staff from one company are assigned to assist with studies from the other company. The people helping out must train on the SOPs from the organization that "owns" the particular protocol. With smaller companies, this is not usually a problem because employees are just given access to both controlled document systems, but in large-company mergers it is not unusual for the employees of one side *not* to have access to the systems on the other side for quite some time—even years! To address this, companies will find that they need to import SOPs from the other "side" into the other system as-is. That is, the SOPs are given new identifiers but otherwise not changed, and this way they can be assigned for training to whoever needs them. This is also the approach when both companies move quickly to the same controlled document system; the SOPs from one side are simply moved into the target controlled document system without changes.

The transition period

For a while, it may be possible to keep studies clearly on one side or the other with their own sets of SOPs. Over time, however—and it may be only a few months for acquisitions of small companies—it is necessary to begin to cut over to a single set of SOPs to be used for all new studies going forward. The single set of SOPs may be ones from either original company, or will be completely new SOPs that reflect the intent of the integrated companies. In this transition period, it will be essential that the SOPs under which a study will be conducted are carefully considered during study start-up and clearly documented in the study's trial master file. Example 2 shows how challenging this can be.

Example 2

In one merger of large companies, the resulting joint company would be using the EDC system of one of the original companies, but under new SOPs, with new procedures designed to reflect best practices for the integrated company. At first, these new SOPs were to be applied to *some* studies started from each side as a kind of pilot. As part of study start-up procedures for these studies, Regulatory Compliance guided the study teams to list all study activities and to specify which SOPs would apply to each of those activities: those of original company 1, company 2, or newly integrated versions. This continued for several months until a full set of integrated SOPs was available that would be applied to all new studies.

The companies in Example 2 did a very good job of creating clear documentation of what was to be followed. Example 3 shows what can happen when that decision is not explicitly made or when the decision is not documented anywhere.

Example 3

A smaller contract research organization (CRO) had been acquired by a larger CRO. During an audit of the smaller unit on behalf of a sponsor that took place over a year after the acquisition of the smaller unit, the auditor was given access to a stack of SOPs and an SOP index. The auditor found the SOPs to adequately address the processes in question; they were in alignment with regulations and regulatory agency expectations and common industry practices. In short, except for some minor points, they were "fine." Training records showed that staff had been trained on the SOPs. The auditor then looked at study documents specific to the studies in scope for the audit and found discrepancies between those documents and the SOPs. Interviews with staff also showed some differences in what was actually being done compared to what was in the SOPs. On bringing this problem up, the auditor was told that the small company was not *actually* following those SOPs that were provided yet (at least on those studies being reviewed); they were still following the "old" ones. Nothing in the SOP materials provided to the auditor indicated that old SOPs were still active and nothing was evident in the study documents provided. Although the staff of the smaller company unit might have been doing exactly the right thing in following the old SOPs, it appeared that no one at either company had carried out the critical step of documenting what was supposed to be followed for each study. There was no document at hand explaining the status of studies from one unit versus the other; the old SOPs were not immediately available in the book of SOPs, and employees were not able to clearly articulate what they should be doing.

Later

For long running studies that are still ongoing long after a merger, a decision has to be made whether to cut them over to new SOPs or keep running under the original ones. For SOPs that are not tightly integrated with software (e.g., amendment of the protocol), this switch can be made. But for others that are tightly tied to the systems they are still using, often the data management SOPs, but sometimes also those for study management and drug safety, this switch is not practical and a way has to be found to keep the old SOPs accessible. The technique mentioned above, to import old SOPs into the designated company-wide controlled document system, is one option. When taking this approach, keep in mind that it may still be necessary to update those transferred SOPs—if nothing else, to update the scope to indicate that the SOP applies to studies on legacy systems only. Unfortunately, a revision might also trigger the requirement to switch to the "new" SOP template.

Controlled Document groups involved in one particular merger of large companies pushed to retire all SOPs that were associated with legacy systems. When it was pointed out that there were still studies running under those controlled documents (in this case, they were studies using paper case report forms), the Controlled Document group suggested that the relevant SOPs simply be filed in the studies' trial master files, but not in the controlled document system. This action would be very dangerous. As staff comes and goes over time, it would be harder and harder to ensure that people newly assigned to the study would be able to *find* the appropriate SOP they had to follow. Documenting training on those SOPs would also be an issue. And, as mentioned above, it might even be necessary to update the legacy SOPs. This latter situation was in fact the case at the company in question, where the SOP guiding paper-based study database locks needed to be updated to support changes in software tools used by those remaining studies. The only way to make this work was to maintain those SOPs for legacy studies in the controlled document system and implement appropriate changes to scope and procedures as soon as they could be revised.

Trial master files

In Chapter 5 we saw how the trial master file (TMF) contents and SOPs are closely tied to each other. During mergers and acquisitions we must not forget that connection. For acquisitions of smaller companies, the transfer of study TMFs should not prove too difficult. The purchasing company should be able to accept the contents in much the same way that a TMF from a fully outsourced study would be transferred from a CRO. For larger companies, and especially mergers, it again becomes more complex. The separate approaches to filing must be maintained for a while. Ongoing studies will typically continue to file with the TMF structure (and system) they started with. For new studies starting up, the Central Records group must provide clear direction as to which TMF applies. Because electronic TMF systems are not fully in use across the industry yet, and because even paper-based TMF systems have different filing models and procedures, staff may have to be trained on *both* TMF models if they are working on multiple studies. Migrating ongoing studies from one TMF to another is such a daunting task, it may not have been done on a large scale by any company yet; typically, studies will close with the TMF they started with, and all new studies will move to any new TMF structure and system.

SOP on acquisitions and mergers?

How to proceed with managing SOPs for mergers will usually not be defined ahead of time and documented in an SOP for mergers or acquisitions. Especially in mergers, both companies are likely to have a say regarding the controlled document system, and the plan for controlled documents should be worked out between them while the companies continue to run independently for a while. Acquisitions are another story as they are becoming a normal business model for growing small and mid-size companies. Successful mid-size companies could well have several acquisitions over just a few years. One example of such a successful and growing mid-size company worked through a few acquisitions and then decided to write down best practices in a document referred to as an "acquisition playbook." The playbook described the general approaches to acquiring locked and ongoing studies and how best to ensure the quality and integrity of the data while upholding good clinical practice. Because the playbook laid out the approaches to assuring data quality, tools to be used, and considerations for integrating the new company, an SOP format was not appropriate—nonetheless, staff involved in acquisitions were expected to follow the guidelines in the playbook. This playbook also addressed how to document the SOPs that would be used for ongoing studies and provided a template project plan. In this use of a playbook, we see again that it is not always necessary to have an SOP in order to get employees to use best practices in insuring the quality of trial data and protection of subjects.

chapter nineteen

Controlled glossaries

Controlled Document groups maintain glossaries to define common terms and abbreviations in a central location so that they do not have to be defined in each controlled document separately. Unfortunately, these company glossaries are often not maintained as well as they could be, leaving SOP readers and authors frustrated when common terms are not included in the glossary and others are included but defined in odd ways. Just as there are good and bad practices to writing and maintaining SOPs, there are also good practices to emulate and other practices to avoid in maintaining glossaries.

Starting the glossary

A company's glossary can be prepopulated with appropriate terms that warrant definition right from the start—especially if the Controlled Document group knows which SOPs are to be created first or if they have a set of initial SOPs already created. Note that some Controlled Document groups do not include terms common to the industry, such as "FDA," in their glossaries, but others do. Even when very common terms are left out, there are other industry terms for which a clear definition in the glossary is valuable; *validation* and *monitoring* are two such examples. Whenever possible, anyone providing a definition should look to the industry for these initial (and also later) terms rather than creating definitions from scratch. Glossaries and definitions from the FDA, EMA,[*] and ICH[†] should always be the first stop, followed by industry professional organizations. Other Internet resources such as Wikipedia can be used—with care—when the definitions found there are associated with reliable references or sources.

After adding necessary industry terms, terms specific to the company's Clinical Development departments can follow. Always include the name used at the company for the study team, as the industry uses a wide variety of terms. Also include the names of, and abbreviations or initialisms for, computer systems used in development. For example, if the electronic data capture system being used is Medidata Rave®, then it is valuable to define *Rave* in the glossary as it is likely to appear in several SOPs;

[*] European Medicines Agency.
[†] International Conference on Harmonisation.

if the safety database is Oracle's Argus Safety, the term defined might be *Argus*. As a general rule, the names of company initiatives are not included even though they are very important and can span a year or more. An initiative's outcomes should find their way into controlled documents without reference to how they got there.

Role names in the glossary

Some companies include key role names that are used in Controlled Documents in the glossary, and this inclusion can have surprising benefits. For SOP authors who want to refer to people performing actions who are outside their own departments, the glossary becomes the place where they can easily access this information. For example, if an author from the Data Management group would like to say that the person in the Clinical Research group in charge of overseeing medical aspects of a trial needs to review adverse event coding, the glossary can clarify if that person is to be called "clinical scientist" or "medical monitor." Keeping roles in the glossary has even more value at companies where roles and job names change. When companies change the names of the main staff roles on study teams—which is quite common in cases of mergers and acquisitions or simple company growth—there is always the question of what to do with the SOPs that use the old name (see Chapter 13). When a role name changes, the glossary can have both the old and the new role name and indicate which supersedes which. Example 1 shows the value of defining roles in the company glossary for a case where a merger resulted in complex study team structures.

Example 1

A large company already had a complex study team structure for international trials where studies typically had many sites. One person led the study team and was known as the global study leader; another acted as a study manager and was known as the global study manager. (The job titles of people filling these two could be the same, they just acted in different capacities depending on the trial.) The study team itself was known as the study management team (SMT). This company then merged with another medium-sized company, which had its own study team structure and role names. That second company would continue as a division of the first and was to run some trials, typically early stage trials, using its own study team structures. The study team at this company was the protocol execution team (PET). The person leading the PET was known as the clinical trial manager; they did not have the study manager role. In writing new SOPs that would apply to all trials conducted in *both* divisions, the company decided to define the roles of study leader and study manager in the company glossary. The study leader was defined as leading the SMT or PET. A global study leader or clinical trial manager filled the role of study leader. The study manager role was also defined and was to be filled by the global study manager of an SMT or the study leader of a PET. That is, for PETs, the clinical trial manager would fill the role of both study leader and study manager. The global SOPs used the study leader and study manager terms to avoid addressing the complexities of who was who in the procedure, and the glossary definition, which used the specifics that applied to each organization, made the SOP clear to staff in both organizations.

If roles are included in the glossary, the standard text found in the *Definitions* section of SOP templates, which usually reads "Refer to the Company Glossary," could be modified to read "Refer to the Company Glossary for terms and role definitions" to remind people to refer to the glossary for roles, also.

Maintaining the glossary

After having defined the initial set of terms, the Controlled Document group can begin to develop guidelines for including terms proposed by staff. SOP authors should always check the company glossary before defining a term in the *Definitions* section of a document. However, they can often forget this step, so when the Controlled Document group reviews the SOP for formatting and consistency with other SOPs, they, too, should check defined terms against the glossary. Any terms that appear in both places should be removed from the SOP if the definitions are easily understood to be the same. If the definitions in the glossary and in the SOP are different, the Controlled Document group and author should confer to decide which is correct and update the glossary as discussed below. The challenge for a Controlled Document group in maintaining the glossary is knowing when to add a term and when to update a term. It may be best to put most of the responsibilities for these decisions on the authors and business process owner (BPO).

A common situation that the Controlled Document group cannot control, but that authors may be able to, occurs when a term is first defined in an SOP and after the SOP is in effect supporting documents are created, such as work instructions. Authors will usually follow the model of the SOP and define the same term in these supporting documents, which opens up the possibility of the definitions diverging over time as reviewers of the work instructions suggest updates to the definition without being aware that it came from an SOP. The best solution for this problem may be for authors of the work instructions to *refer* to the definitions in the associated SOP rather than to *copy* the definitions.

This brings up the question of who decides if a proposed definition is correct. Terms in the glossary will impact multiple documents and so probably should go through a review and approval process similar to that for a document—even if the definition is initially taken from an SOP that has been reviewed. In practice, definitions in SOPs often do not get a really thorough review because the reviewers are generally familiar with the term in the first place. One good approach is to circulate any definition that is a candidate for inclusion in the glossary through a review cycle by itself, without the rest of the document. The result of that review should then go to the Controlled Document group for review or to a special set of reviewers for glossary terms. The Controlled Document group should include a check, through text searches, of other controlled documents for that term. Approval of the definition by someone from the Controlled Document group, indicating that the process of review has been followed, should be enough, but some companies may choose to have a glossary owner or a BPO approve new terms.

Note that following a procedure to add a term to the glossary when it is needed in a second document means that it will still appear in the *Definitions* section in the first controlled document, which is still in effect. As long as the terms are well aligned and the same in intent, it should not cause a problem. SOP authors often know that a term will be needed in multiple documents and should be permitted to propose an addition to the glossary proactively.

Maintaining glossary terms by updating or removing them will be easier if there is some information about where the term came from originally; though that information would not be published or visible in the glossary that users access, it would be still associated with each term. The Controlled Document group could record the following for each term: a date the term was added, the documents it appeared in at the time it was added, the source of the definition, and the group proposing the definition. (It would not be necessary, or even practical, to list all the documents in which it is used over time.) If the glossary is posted in the form of a PDF document, the Controlled Document group can create the PDF from a subset of the source table of information at each revision. Later, when the term is to be updated or removed, the original group can be consulted.

Updating a definition of a term is, of course, risky because the controlled documents in effect might have used the original version of the term. Except for slight updates for clarity, updating of a definition should only be done with great caution and in consultation with the authors or BPO of the original document. Removing terms from the glossary is less risky but to avoid confusion, they should only be removed when they are no longer used. Unfortunately we cannot easily tell when revisions to documents result in terms no longer being used, however, the Controlled

Document group can request a review of all terms in the glossary every few years by original requestors. If an originator recommends removing the term, a full text search of controlled documents should still be performed to confirm that the term, or its definition, is truly no longer required.

Using glossaries

As mentioned in Chapter 9, having a hyperlink to the glossary in the SOP's *Definitions* section is a very good way to encourage readers to actually look something up. The next best option, and a good backup in any case, is to have a link to the glossary directly on the same web page as is used to access the controlled document library. The Controlled Document group should encourage departments that create internal web pages for access to their commonly used SOPs to also include a link to the glossary on that page. The glossary is unlikely to be used on a regular basis if a reader has to go to the controlled document library and do a search to see the definition of a term.

Appendix one: Example SOP template

This example SOP template reflects many of the recommendations found in Chapter 9. A template often has instructions associated with it. In this example, the instructional document is referred to as CDOC-902-MAN, "SOP Author Style Guide," following the numbering scheme introduced in Chapter 14.

Document ID: BPO-NNN-SOP	Version: 1.0	Effective: DD-MO-YYYY
Document Title: SOP-title-here		PAGE 1 of 2

1 Purpose

Expand on the title to say what *activities* will be described. It will sometimes be appropriate to explicitly state what is *not* covered in the procedure.

2 Scope

Say which *trials* or under what conditions this procedure will apply to. Be specific but allow flexibility if there is any question.

3 Responsibility

List the roles that have responsibilities in this SOP. For roles defined for the purposes of this SOP, provide the list of job titles that may take on that role.

3.1   The following roles have active responsibilities for completion of tasks:

3.2   The following roles only review or approve documents or actions:

4 Definitions

See also: Clinical Development Glossary

Define any additional terms not already present in the glossary. Refer to CDOC-902-MAN, "SOP Author Style Guide," and consult with the CDOC group for additional guidelines.

5 Background

Use this optional section to provide context for the process that a reader would need to fully understand the process as a whole. Use a brief introduction to the procedures in Section 6 as an alternative.

6 Procedure

Divide the procedure into subprocedures as appropriate.

6.1   Subprocedure

[Optional] Brief context for this part of the procedure. The example table below shows the verb form for single and multiple roles and also uses the approach of listing review and approvals as separate rows to highlight responsibility.

No.	Responsible	Procedure
6.1.1	Role 1	Perform the first task and sends it to Role 2 and Role 3 for review and approval.
6.1.2	Role 2 and Role 3	Review and approve the document.
6.1.3	Role 1	Continue with the next step.
6.1.4		

Document ID: BPO-NNN-SOP	Version: 1.0	Effective: DD-MO-YYYY
Document Title: SOP-title-here		PAGE 2 of 2

6.2 Subprocedure

No.	Responsible	Procedure
Prerequisite	Before proceeding to the steps that follow, a particular condition must first be met or a procedure governed by another SOP has to be complete.	
6.2.1	Role 2	Continue once the prerequisite has been met.
6.2.2		

7 Document disposition

For *each* kind of document or output (form, document, listing, report, etc.) created or used in the procedure, specify its disposition.

Document or output	Disposition
[provide name and ID, if applicable]	Choose one: TMF, Shared Folders, Not Retained, or Other. If Not Retained or Other, explain.

8 References

List only those controlled documents and department-managed documents specifically mentioned in the procedure. Also include all forms and templates used.

9 Appendices

List appendices here. Appendices are optional but a process map is a recommended. Each appendix must be referred to in at least one of the previous sections. Begin appendices after a page break. Use a title style for the name of each appendix.

Revision history

Version	Effective date	Author	Changes
01	DD-MO-YYYY	B. Author	Updated Section 6.1 to .... Corrected 6.2.1 to read ....
00	DD-MO-YYYY	A. Author	Original

Appendix two: Example SOP of SOPs

The example SOP of SOPs below reflects many of the recommendations found in the chapters of this book and in the example SOP template found in Appendix 1. The example reflects the following assumed attributes of an imagined system for controlled documents:

- The terms *SOP* and *controlled document* are defined in the corporate glossary.
- Deviations and requirements for training on controlled documents are covered in a separate SOP.
- Several manuals in the controlled document system would support the SOP. They contain detailed information that is inappropriate for inclusion in the SOP. These use the numbering scheme introduced in Chapter 14.
- Another SOP governs compliance review; it is explicitly referenced here to ensure that there is a link between the two documents. If the SOP on compliance review is ever updated, this link would ensure that the SOP of SOPs would be assessed to see if a change in the tasks or their order was required.
- The SOP's Appendix 1 would contain the company's version of the document hierarchy, which might be a version of the diagram found in Chapter 2, Figure 2.2.

Note that the *Responsibility* section in this SOP is not divided out into the roles with active responsibility and roles that review and approve for this SOP as the template recommends. This choice reflects the reality that anyone in Clinical Development could act as the business process owner representative or as a reviewer of an SOP during any of the rounds of review required for a controlled document governed by this procedure. In this case, everyone in Clinical Development would be called upon to train on this SOP, and that is in fact the most common approach for the SOP of SOPs.

Document ID: CDOC-001-SOP	Version: 1.0	Effective: DD-MO-YYYY
Document Title: Managing Controlled Documents		Page 1 of 6

1 Purpose

This SOP describes the activities for initiating, identifying, writing, approving, posting, updating, and retiring controlled documents that apply to Clinical Development departments.

2 Scope

This procedure applies to management of all controlled documents intended to apply to CORP Clinical Development activities. Deviations from SOPs are addressed in CDOC-003-SOP, *Recording Prospective and Retrospective Controlled Document Deviations*. For documents maintained by departments to support controlled documents, refer to CDOC-901-MAN, *Guidance for Maintaining Department-Managed Documents*.

3 Responsibilities

Anyone in Clinical Development may act in the role of business process owner representative or participate in the review of an SOP. Management may be asked to approve an SOP as defined in Appendix 2.

4 Definitions

See also: Clinical Development Glossary

Administrative change: a change to *information* in a controlled document to make a correction that does not change any procedure or responsibility. Approval is by a Controlled Document group manager. No training is required for administrative changes.

Business process owner (BPO): the department within Clinical Development that takes responsibility for a given controlled document. Usually, the BPO is the group most heavily involved in, or impacted by, the document.

Major change: a change to a controlled document that impacts the activities, the order of the activities, or responsibilities. See approval requirements in Appendix 2. Training is always required for major changes.

Minor change: a change to a controlled document to clarify the procedure but not change it. Approval is by the BPO representative who initiated the change. No training is required for minor changes.

5 Background

At CORP, in addition to SOPs, the following types of controlled documents are supported: forms, templates, work instructions, and manuals. Refer to Appendix 1 for additional information.

6 Procedure

6.1 Creating controlled documents

The group initiating a new SOP is generally designated as the BPO. When a process is heavily cross-functional, the groups most impacted by the procedures decide on a BPO.

Document ID: CDOC-001-SOP	Version: 1.0	Effective: DD-MO-YYYY
Document Title: Managing Controlled Documents		Page 2 of 6

No.	Responsible	Procedure
6.1.1	BPO Representative	Fill out a change request form (CDOC-001-FRM-1) for the new document(s). A tentative title, purpose, and target date are required. Forward the form to the Controlled Document Coordinator.
6.1.2	Controlled Document Coordinator (CDOC)	Assign a controlled document identifier and provide it to the BPO representative.
Prerequisite		Before drafting the document, the BPO must ensure that the process is ready to be used on clinical trials and related activities. Refer to CDOC-900-MAN, "Guidelines for Process Development."
6.1.3	BPO Representative	From the process, draft the document using the appropriate controlled document template (CDOC-001-TMP-1 through CDOC-001-TMP-4). Refer to CDOC-902-MAN, "SOP Author Style Guide," for instructions on using the templates. Proceed with Section 6.3 when the draft is ready for review.

6.2   Revising controlled documents

All Clinical Development staff are expected to notify the BPO or CDOC when they recognize that updates to controlled documents are necessary and should refer to CDOC-003-SOP, "Recording Prospective and Retrospective Controlled Document Deviations," to determine whether a formal deviation is necessary.

No.	Responsible	Procedure
6.2.1	BPO Representative	Fill out a change request form (CDOC-001-FRM-1) when working on a revision. Forward the form to the CDOC.
6.2.2	CDOC	If the change is classified as an administrative change (see "Definitions"), make the change in consultation with the BPO and post the document as effective. Otherwise, provide the editable copy of the document to the BPO.
6.2.3	BPO Representative	In the agreed-upon timeframe, update the controlled document and proceed with Section 6.3.

6.3   Reviewing controlled documents

BPOs are responsible for identifying reviewers within their departments. CDOC maintains a list of contacts in each department or function to serve as a point of contact for cross-functional reviews. At any point in the review cycles, if feedback results in substantive changes the responsible role should send the revised document back for a full re-review to all reviewers, or to select reviewers with particular interest in the changes.

Document ID: CDOC-001-SOP	Version: 1.0	Effective: DD-MO-YYYY
Document Title: Managing Controlled Documents		Page 3 of 6

No.	Responsible	Procedure
6.3.1	BPO Representative	Send the draft for review by appropriate subject matter experts. Incorporate feedback.
6.3.2	BPO Representative	Send the document for review by BPO reviewers. Incorporate feedback.
6.3.3	BPO Representative	Send the draft to CDOC.
6.3.4	CDOC	If cross-functional review is required, send the draft to each department that has a role in the procedure, including those that have review and approval activities. Ensure the BPO representative receives all feedback.
6.3.5	BPO Representative	Together with subject matter experts, review and incorporate feedback as appropriate. Provide CDOC with the updated document(s).
6.3.6	CDOC	For revisions, identify all controlled documents that refer to the document being revised and request that the BPOs of those documents assess the impact of the changes. If updates to other documents are needed but cannot be completed together, request that those BPOs file a planned deviation to each impacted SOP.
6.3.7	CDOC	If compliance review is required (refer to Appendix 2), send the document for compliance review, which is carried out according to RC-205-SOP, "Conducting Compliance Review of Controlled Documents."
6.3.8	CDOC	In parallel to compliance review • Send document to TMF Governance for comment if applicable (refer to Appendix 2). • Identify expected approvers (see Section 6.4) and send a copy of the document to all approvers for a courtesy review.
6.3.9	BPO Representative	Incorporate any feedback from compliance and approver review in consultation with the reviewers and subject matter experts. Provide the final draft to CDOC.
6.3.10	CDOC	Assess whether the changes made by the BPO are classified as a major or minor change to the controlled document and continue with 6.4.

6.4   Approving and posting controlled documents

CDOC maintains the list of approvers for each department and ensures that they are trained in the approval process (see also Appendix 2).

Document ID: CDOC-001-SOP	Version: 1.0	Effective: DD-MO-YYYY
Document Title: Managing Controlled Documents		Page 4 of 6

No.	Responsible	Procedure
6.4.1	CDOC	Review the document to ensure proper formatting and consistency with other controlled documents. Consult with the BPO as necessary.
6.4.2	BPO Representative	Provide CDOC with cut-over rules to be documented in the implementation memo when the SOP is posted as effective.
6.4.3	CDOC	Send document for approval. When all approvals have been obtained, route the document for posting. Notify the BPO representative of the date the document will be prereleased for training. Consult with the BPO representative on a mutually agreeable effective date for the document.
6.4.4	CDOC	Follow CDOC-002-SOP, "Training for Controlled Documents," to prepare for posting for training. Notify the BPO representative when the document posts.
6.4.5	CDOC	On the agreed-upon effective date, notify the BPO representative that the controlled document is now effective.
6.4.6	CDOC	Update the controlled document index.

6.5   Retiring controlled documents

After a controlled document is retired, it can only be accessed by the Controlled Document group.

No.	Responsible	Procedure
6.5.1	BPO Representative	Request retirement of a controlled document using CDOC-001-FRM-1.
6.5.2	CDOC	Identify all controlled documents that refer to the document to be retired.
6.5.3	CDOC and BPO Representative	Perform an initial assessment of how the impacted documents would have to be changed.
6.5.4	CDOC	If any impacted documents belong to other BPOs, notify those BPO representatives of the proposed retirement and recommended updates.
6.5.5	CDOC	If a BPO cannot complete updates to an impacted document by the time the triggering document retires, request that the BPO file a prospective deviation for that document.
6.5.6	CDOC	Request approval for the retirement according to Appendix 2. When all approvals have been obtained, consult with the BPO representative to set a retirement date.
6.5.7	CDOC	On the agreed-upon date, notify the BPO representative that the controlled document is now retired.

Document ID: CDOC-001-SOP	Version: 1.0	Effective: DD-MO-YYYY
Document Title: Managing Controlled Documents		Page 5 of 6

6.6   Periodic review of controlled documents

SOPs and work instructions must be reviewed *and updated* every two years.

No.	Responsible	Procedure
6.6.1	CDOC	Notify the BPO representative three months before a controlled document reaches its two-year anniversary. Provide an editable copy of the document to the BPO representative.
6.6.2	BPO Representative	File a prospective SOP deviation if the SOP cannot be reviewed, updated, and approved within one month *past* the anniversary date.
6.6.3	BPO Representative	Route the document for thorough review according to Section 6.3.
6.6.4	CDOC	Even if the proposed changes are technically administrative changes, proceed as for a minor change. Follow Sections 6.3 and 6.4.

7   Document disposition

Document or Output	Disposition
All approved controlled documents	GCP-Doc*
CDOC-001-FRM-1 "Controlled Document Change Request Form"	GCP-Doc

*GCP-Doc is CORP's validated, controlled document system.

8   References

CDOC-001-FRM-1 "Controlled Document Change Request"
CDOC-001-TMP-1 through 4, "Templates for Controlled Documents"
CDOC-002-SOP, "Training for Controlled Documents"
CDOC-003-SOP, "Recording Prospective and Retrospective Controlled Document Deviations"
CDOC-900-MAN, "Guidelines for Process Development"
CDOC-901-MAN, "Guidance for Maintaining Department-Managed Documents"
CDOC-902-MAN, "SOP Author Style Guide"
RC-205-SOP, "Conducting Compliance Review of Controlled Documents"

9   Appendices

Appendix 1: Controlled Document Types
Appendix 2: Characteristics of Controlled Documents

Document ID: CDOC-001-SOP	Version: 1.0	Effective: DD-MO-YYYY
Document Title: Managing Controlled Documents		Page 6 of 6

Appendix 1: Controlled document types

Work instructions: These detailed procedures describe how activities in the parent SOP are to be carried out; work instructions are limited to a single department and must be associated with an SOP.

Forms: These provide a highly structured set of information needed to carry out a process or to document that an activity has taken place; a form must be associated with an SOP or work instruction.

Templates: These provide a structure and (optional) default content for a document that is then customized with information specific to a study or activity in question; templates must be associated with an SOP or work instruction.

Manuals: These documents provide additional information or instructions for regulated or business critical activities; the content of manuals may take many forms including but not limited to handbook, guide, best practices document, and playbook. Manuals do not have to be associated with an SOP. Manuals are associated with a business process owner but can apply to multiple groups within that BPO.

Appendix 2: Characteristics of controlled documents

	Application area	Compliance review?	Document disposition review?	Type of approval (major change)	Supports
SOP	One or more departments in Clinical Development	Yes	Yes	Head of each department with responsibilities	N/A
Work Instruction	One department in Clinical Development	Yes	Yes	Head of department	SOP
Form	Same as document	No	No	Head of BPO*	SOP or work instruction
Template	Same as document	No	No	Head of BPO*	SOP or work instruction
Manual	Single business process owner	No	No	Senior manager or above in BPO	Stands alone

*If updated independently of the parent document.

Revision history

Version	Effective date	Author	Changes
00	DD-MO-YYYY	A. Author	Original

Index